Time in History

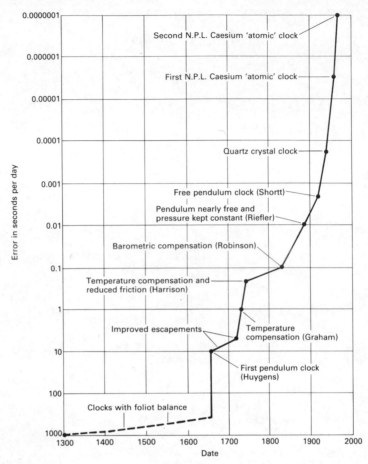

A graph illustrating the increasing precision of timekeeping. This graph is based on the chart devised by F.A.B. Ward, formerly of the Science Museum, London. It illustrates the increasing rate of increase in the accuracy of timekeeping that has occurred since the invention of the first mechanical clock about AD 1300. It has been estimated by Joseph Needham that the accuracy of Su Sung's Chinese water-clock of about AD 1100 was such that, if it were represented by a point on this graph, its vertical ordinate would be between 10 and 100. Two horologists whose names occur on this graph are not mentioned in the main text: Robinson and Riefler. Thomas Robinson (1792–1882) was an Irish astronomer who became Director of the Armagh Observatory. In 1831 he attached a small mercury barometer to a clock-pendulum in order to make it possible to compensate for the rather complicated effect of barometric pressure on the pendulum's rate of oscillation. In 1889 A. Riefler of Munich patented an escapement in which the impulses of the pendulum were transmitted through its suspension spring, the pendulum being otherwise free from interference. His clocks were so successful that they were adopted as standard clocks in many observatories until they were replaced by Shortt clocks.

Time in History

The evolution of our general awareness of
time and temporal perspective

G. J. WHITROW

Oxford   New York
OXFORD UNIVERSITY PRESS

70611337

Oxford University Press, Walton Street, Oxford OX2 6DP

Oxford New York Toronto
Delhi Bombay Calcutta Madras Karachi
Petaling Jaya Singapore Hong Kong Tokyo
Nairobi Dar es Salaam Cape Town
Melbourne Auckland

and associated companies in
Berlin Ibadan

Oxford is a trade mark of Oxford University Press

British Library Cataloguing in Publication Data
Whitrow, G. J. (Gerald James),
Time in history: the evolution of our general awareness
of time and temporal perspective.
1. Time. Cultural aspects, to 1987
I. Title
529'.09
ISBN 0–19–215361–7

Library of Congress Cataloging in Publication Data
Whitrow, G. J.
Time in history: the evolution of our general awareness
of time and temporal perspective/G. J. Whitrow.
p. cm. Bibliography: p. Includes index.
1. Word history. 2. Time—Social aspects. 3. Time
perception. 4. Chronology. I. Title.
909—dc19 D21.3.W47 1988 88–4245
ISBN 0–19–215361–7

Printed and bound in Great Britain by
Biddles Ltd, Guildford and King's Lynn

To Magda

Preface

Most of us are so accustomed to the ideas of time, history, and evolution that we are inclined to forget that these concepts have not always been accorded the importance which we now assign to them. If, however, we are to understand why it is that time tends to dominate our way of life and thought, we must acquire some knowledge of how this has come about. In other words, we must put time itself into temporal perspective. The purpose of this book is to present the main features of the evolution of our general awareness of time and its significance in a form suitable for all who are interested in the subject.

The present volume can be regarded as supplementing my book *The Natural Philosophy of Time*,[1] the second edition of which was published by the Clarendon Press in 1980. The publication of the first edition in 1961 gave rise to a considerable increase in interest in the general scientific study of time (as distinct from that of temporal logic, on the one hand, and horology, on the other—topics which my book did not cover), and led, on the initiative of J. T. Fraser, to the formation of the International Society for the Study of Time, the first Conference of which was held under my Presidency, with Dr Fraser as Hon. Secretary, at Oberwolfach, West Germany, in 1969. Nevertheless, only two books have so far been devoted to the *history* of our awareness of time and its significance, namely *The Discovery of Time*, by Stephen Toulmin and June Goodfield,[2] and the more comprehensive *Zeit und Kultur: Geschichte des Zeitbewusstseins in Europa*, by Rudolf Wendorff;[3] the contribution of Howard Trivers should also be mentioned.[4] Moreover, these books, admirable as they are, have been written from the point of view of general intellectual history, whereas I have taken more account of the developments that have occurred in chronology and chronometry and their social and ideological consequences.

Besides discussing the general influence of time on the mental outlook and way of life in different ages and civilizations, I have directed particular attention to the history of the measurement of time. The crucial stage in this development was the invention of the mechanical clock in Western Europe towards the end of the thirteenth century. For, despite the mystery which still surrounds the event, its consequences were far-reaching and ultimately led to the dominating role of time in contemporary civilization.

I should like to thank the publishers Thames & Hudson, of London, and the Plenum Press, New York, respectively, for kindly allowing the present book to overlap, in a few passages, with my book *What is Time?* (1972) and my chapter 'The Role of Time in Cosmology' in the book edited by W. Yourgrau and A. D. Breck, *Cosmology, History and Theology* (1977).

I am grateful to Imperial College for appointing me to a Senior Research Fellowship on my retirement, thereby allowing me to continue to make full use of the facilities of the College. My thanks are due in particular to Jagna Pindelska, Librarian of the Mathematics Department, Imperial College, for sparing no pains in tracking down any book or article that I wished to consult. I also owe a great debt to my wife, Magda Whitrow, who has not only read the manuscript, but has produced both the typescript and the Index with her customary skill.

January 1988   G. J. W.

Contents

Man, in a word, has no nature; what he has is . . . history.

J. Ortega y Gasset ('History as a System'. In *Philosophy and History: the Ernst Cassirer Festschrift*)

To ingenious attempts at explaining by the light of reason things which want the light of history to show their meaning, much of the learned nonsense of the world has indeed been due.

Edward B. Tylor (*Primitive Culture*)

Perhaps the most important way the urban bourgeoisie spread its culture was the revolution it effected in the mental categories of medieval man. The most spectacular of these revolutions, without a doubt, was the one that concerned the concept and measurement of time.

J. Le Goff (*The Fontana Economic History of Europe: the Middle Ages*)

And he that will not apply New Remedies, must expect New Evils: for Time is the greatest *Innovateur*.

Francis Bacon (*Essay XXIII*: 'Of Innovations')

The point is that in the past the time-span of important change was considerably longer than that of a single human life. Thus mankind was trained to adapt itself to fixed conditions.

To-day the time-span is considerably shorter than that of human life, and accordingly our training must prepare individuals to face a novelty of conditions.

A. N. Whitehead (*Adventures of Ideas*)

Part I
Introduction

1. Awareness of Time

Time and civil life

Most of us feel intuitively that time goes on forever of its own accord, completely unaffected by anything else, so that if all activity were suddenly to cease time would still continue without any interruption. For many people the way in which we measure time by the clock and the calendar is absolute, and by some it has even been thought that to tamper with either was to court disaster. When, in 1916, Summer Time was first introduced in the United Kingdom by advancing the clock one hour, there were many who objected to interfering with what the popular novelist Marie Corelli called 'God's own time'. Similarly, in 1752, when the British government decided to alter the calendar, so as to bring it into line with that previously adopted by most other countries of Western Europe, and decreed that the day following 2 September should be styled 14 September, many people thought that their lives were being shortened thereby. Some workers actually believed that they were going to lose eleven days' pay. So they rioted and demanded 'Give us back our eleven days!' (The Act of Parliament had, in fact, been carefully worded so as prevent any injustice in the payment of rents, interest, etc.) The rioting was worst in Bristol, in those days the second largest city in England, where several people were killed.

Even today, when we are all familiar with the idea of altering the time on the clock so as to suit our general convenience, it still comes to many of us as something of a shock when we are first made to realize that there is, for example, a five-hour difference between London and New York, so that when it is ten o'clock in the evening—nearly bedtime, in London—it is only five o'clock in the afternoon in New York. Moreover, even the most experienced and sophisticated among us can suffer the peculiar, and often unpleasant, effects of 'jet-lag' when we fly a long way in an easterly or westerly direction. No less strange, although unaccompanied by any peculiar physiological symptoms, is the effect of crossing the International Date Line, which has been drawn in a zigzag fashion down the Pacific from one pole to the other. For, when a ship or plane on its way from, say, San Francisco to Hong Kong crosses this line it loses a whole day of the calendar because of the time difference of twenty-four hours between any position immediately to the east of the line and any position to the west of it. Although, in this case, there is no need to adjust our watches, we have to discard a day from the week concerned. On the other hand, when we cross the line in the opposite direction we appear to experience an eight-day week, so that if the crossing is made precisely at midnight we live through two Fridays, say, in succession. This means that in going round the world eastwards the number of days occupied on the journey will be one more than the number of days reckoned at the point where the journey begins and ends, each day on the journey being less than twenty-four hours, but if the journey is made westwards the number of days taken will be one less than the number reckoned at the point of commencement, each day on the journey being longer than twenty-four hours. This phenomenon was made the basis of the story *Around the World in Eighty Days* by Jules Verne, in which after the hero had completed his journey eastwards he thought he had taken over eighty days, but since he omitted to put his calendar back when crossing the Date Line he found on his return that he was one day ahead of the calendar and so he had, after all, completed his journey in the prescribed time.

All these experiences seem strange because they appear to conflict with our intuitive feeling that time is something universal and absolute. What gives rise to these phenomena is the way we choose to measure time and relate it to the way we live. The time kept by us in civil life is based on the rotation of the earth, which gives us our day. Similarly, the earth's motion around the sun gives us our year. If, however, we lived on the moon, we should then find that, since the moon spins on its axis so much more slowly than does the earth, each day as determined by the moon's rotation would in fact be equal to a month. The way in which the terrestrial day is divided up into hours, minutes, and seconds is purely conventional. Similarly the decision whether a given day begins at dawn, sunrise, midday, sunset, or midnight is also a matter of arbitrary choice or social convenience.

Our sense of time

Granted, then, that the time of civil life is measured in a way that happens to suit us on earth but has no absolute or universal significance, what about our inner feeling of time? Is it this that provides us with our intuition of the absolute nature of time? Time certainly is a fundamental characteristic of human experience, but there is no evidence that we have a special sense of time, as we have of sight, hearing, touch, taste, or smell. Our direct experience of time is always of the present, and our idea of time comes from reflecting on this experience. Nevertheless, so long as our attention is concentrated on the present we tend to be unaware of time. A 'sense of time' involves some feeling or awareness of duration, but this depends on our interests and the way in which we focus our attention. If what we are doing interests us, then time seems short, but the more attention we pay to time itself, that is to its duration, the longer it seems. Never does a minute seem so long as when we look at the seconds hand moving round the face of a watch or clock. Clearly, then, our belief in the absolute nature of temporal duration is not an immediate consequence of our experience, but as I have just said comes from reflecting on this experience. Our sense of duration is affected not only by the degree to which we concentrate our attention on what we are doing but by our general physical condition. In particular, it can be distorted by drugs or by our being confined for long periods to cold, dark environments and being deprived of clocks and watches. But the most widely experienced factor that influences our sense of duration is our age, for it is generally recognized that, as we get older, time as registered by the clock and the calendar appears to pass ever more rapidly.

We experience a feeling of duration whenever the present situation is related by us either to our past experiences or to our future expectations and desires. There is no evidence that we are born with any sense of temporal awareness, but our sense of expectation develops before our consciousness of memory. When a very young child cries with hunger he has his first experience of duration, but these temporal experiences are isolated. It has been suggested that the relatively long delay experienced by the young child in acquiring the ability to walk has an important influence on the development of our sense of time, since the child's eagerness to grasp what he cannot reach gives rise to the first primitive notion of time, associated with a space that cannot be crossed.[1] Even when the child begins to walk, to reach is also still to wait and hence enhances the feeling of delay associated with expectation. The first intuition of duration appears as an interval which stands between the child and the fulfilment of his desires.

The child's gradual acquisition of temporal concepts can be closely correlated with the development of his use of language. For, although our awareness of time is a product of human evolution, our ideas of time are neither innate nor automatically learned but are intellectual constructions that result from experience and action.[2] Up to the age of 18 months or more children appear to live only in the present, and by that age the meaning of 'now' has usually been acquired. Between then and 30 months, although most of the time-related words that children learn to use deal only with the present, they tend to acquire a few words relating to the future, such as 'soon', but almost none that concern the past. Consequently, the use of 'tomorrow' precedes that of 'yesterday', although at first both are likely to be interpreted as meaning 'not today'. As the child grows older, the relative proportion of present-oriented statements tends to decrease but still predominates, future-oriented statements increase somewhat, but past-oriented statements increase more slowly. Nevertheless, young children have difficulty in acquiring a unified concept of time, for even when the child begins to recognize temporal sequences time remains dependent on his own activities. The gradual acquisition of language, however, not only increases the child's ability to understand and communicate but also enables him to grasp temporal relationships and to extend his ability for temporal conceptualization. For, although awareness of temporal phenomena may seem to be inherent in our personal experience, it involves an abstract conceptual framework which we only gradually learn to construct.[3] Even when the child begins to associate time with particular external movements, he is not truly conscious of time until he begins to realize that things bear a relationship not only to each other but also to himself, and this only becomes possible with the development of memory. The child's sense of memory involves not only events in his own experience but, in due course, some in the memory of his parents and eventually events in the history of his social group. It is not until about the age of 8 or later that the relations of temporal order (before and after) are associated with duration so as to lead to the idea of a single common time in which all events happen. It has been found that at the age of 10 only one child in four regards time as an abstract concept independent of actual clocks. Not surprisingly, the ability to grasp this idea depends on the rate of development of the child's intelligence. An experiment performed on children between the ages of 10 and 15, to test whether they thought they had become older when clocks were advanced one hour to 'Summer Time', revealed that at the younger age only one child in four believed that this change of time had had no effect on his age. Only when they are 13 or 14 do most children begin to realize that the time indicated on a clock is a convention.[4]

This description of the way in which children learn to develop their sense of time applies only to those growing up in Western industrial civilization and not to children in less sophisticated societies. For example, P. M. Bell has reported that when teaching children in Uganda he found that, although they were not unintelligent, they had much greater difficulty than Western children of similar age in judging how long something took to happen, a two-hour journey by bus being said by some to have taken only ten minutes and by others six hours![5] Also, Australian aborigine children of similar mental capacity to white children find it extremely difficult to tell the time by the clock—something that most Western children usually have learned to do successfully by the age of about 6 or 7. The aborigine children can read the hands of the clock as a memory exercise, but they find it difficult to relate the time they read on the clock to the actual time of day. The explanation that has been suggested is that their lives, unlike ours, are not dominated by time.[6]

Time and mankind

Our sense of time involves some awareness of duration and also of the differences between past, present, and future. There is evidence that our sense of these distinctions is one of the most important mental faculties distinguishing man from all other living creatures. For we have good reason to believe that all animals except man live in a continual present. The possession by animals of some sense of memory, as shown, for example, by dogs which are inclined to give vent to the wildest joy on seeing their owners after a long separation, does not necessitate any image of the past as such. It is sufficient for the dog to recognize its owner.

Similarly, there is no firm evidence that animals have any sense of the future. In general any actions of theirs that might be thought to bear on this question seem to be purely instinctive, although this conclusion is not quite so obvious in the case of the higher apes, particularly the chimpanzee. The problem was considered very carefully by Wolfgang Koehler in the course of his famous investigation of the mentality of apes. He studied cases where chimpanzees undertook, with a view to some final goal, preparatory work that lasted a long time and in itself afforded no visible approach to the desired end. In such cases it seemed at first that the animal might have some rudimentary notion of the future. Nevertheless, Koehler came to the conclusion that all such behaviour by the highest apes could be explained, to quote his own words, 'more directly from a consideration of the present only'.[7] In particular, after a careful analysis of experiments in which chimpanzees readily responded to the opportunity given them to postpone eating until they had accumulated a large supply of food to eat later in some quiet corner free from disturbance, Koehler could find no reason for interpreting their conduct as evidence for a sense of the future. Instead of the animal being spurred on by some feeling of what it will be like later when eating the food, he believed that the chimpanzee's behaviour was simply a response to its instinctive desire to get as much food as possible now. More recently, in his detailed monograph *Animal Thought*, Stephen Walker has confirmed Koehler's views and has concluded that it is surprisingly difficult to produce convincing experimental proof that any animal has any memory or foresight at all.[8]

Nevertheless, the conclusion that a sense of time is peculiar to mankind needs careful evaluation. For, whereas in the absence of any incontrovertible counter-evidence we have good reason to deny this faculty to animals, it has been claimed that there are human beings who also manage very well without it. The classic example that has often been cited is that of the Hopi of Arizona, whose language was studied in great detail by Benjamin Lee Whorf.[9] He concluded that the Hopi language contains no words, grammatical forms, constructions, or expressions that refer to time or any of its aspects. Instead of the concepts of space and time the Hopi use two other basic states which Whorf denoted by the terms 'objective' and 'subjective', respectively. The objective state comprises all that is or has been accessible to the senses, with no distinction being made between present and past, although everything that we call future is excluded. The subjective state comprises all that we would regard as mental or spiritual, including everything that for us is future, much of which the Hopi regard as predestined, at least in essence. It also includes an aspect of the present, namely that which is beginning to be revealed or done, for example starting an action such as going to sleep. The objective state includes all intervals and distances and in particular the temporal relations between events that have already happened. The subjective state, on the other hand, comprises nothing corrresponding to the sequences and successions that we find in the objective state. Unlike English, the Hopi language prefers verbs to nouns, but its verbs have no tenses. The Hopi do not need terms that refer to space or time. Terms that for English-speakers refer to these concepts are replaced by expressions concerning extension, operation, and cyclic process if they refer to the objective realm. Terms that refer to the future, the psychic-mental, the mythical, and the conjectural are replaced by expressions of subjectivity. Whorf claims that, as a result, the Hopi language gets along perfectly without tenses for its verbs.

Whorf's contention that 'the Hopi language contains no reference to 'time', either explicit or implicit, is, however, too sweeping.[10] For there is a temporal distinction between the two basic forms of Hopi thought. Instead of the three temporal states—past, present, and future—the Hopi imagine two states which between them comprise our past, present, and future. In so far as the Hopi recognize implicitly a distinction between past and future, they cannot be said to live only in the present. They have some sense of time, although their fundamental intuition of time is not the same as the one evolved in Europe. Nevertheless, the Hopi have successfully developed an agricultural and ceremonial calendar, based on astronomical lore, that is sufficiently precise for particular festivals seldom to fall more than two days from the norm.[11]

Similarly, as has been pointed out by Evans-Pritchard, time has a different significance for the Azande of southern Sudan than it had for him. From their behaviour he concluded that for them present and future overlap, so that a man's future health and happiness depend on future conditions that are regarded as already existing. Consequently, it is believed that the mystical forces which produce these conditions can be tackled here and now. When the oracles indicate that a man will fall ill in the near future his state is already bad, his future being already a part of present time. Although the Azande cannot explain these matters, they are content to believe them and act upon them.[12]

Another Sudanese race studied by Evans-Pritchard, the Nuer, who live on both banks of the White Nile, have no equivalent of our word 'time' and cannot speak of it as if it were something that passes and can be saved or wasted. Their points of reference in time are provided by their social activities. 'Events [for them] follow a logical order, but they are not controlled by an abstract system, there being no autonomous points of reference to which activities have to correspond with precision.'[13] The Nuer have no units of time such as hours or minutes, for they do not measure time but think only in terms of successions of activities. So many of these involve their cattle that Evans-Pritchard speaks of their 'cattle-clock'. Years are referred to by the floods, pestilences, famines, wars, and so on occurring in them. In due course the names given to the years are forgotten and all events beyond the range of this crude historical record come to be regarded as having occurred long ago. Historical time based on a sequence of events that are of great significance for a whole tribe goes back further than the historical time of smaller groups, but in

Evans-Pritchard's opinion it never covers a period of more than about fifty years and the further back from the present the fewer and vaguer are its points of reference.[14] Distance between events is not reckoned by the Nuer in terms of temporal concepts but in terms relating to social structure, notably what Evans-Pritchard calls the 'age-set system', all boys 'initiated' during a number of successive years belonging to a single age-set. At the time of Evans-Pritchard's investigation of this system he found members of six sets alive. Although he was unable fully to elucidate the way in which an individual actually perceives time, since the subject 'bristles with difficulties', he concluded that for the Nuer the perception of time is no more than the movement of persons, often as groups, through the social structure. Consequently, it does not yield a true impression of temporal distances between events like that produced by our techniques of dating. In particular, the temporal distance between the beginning of the world and the present day remains fixed. Time-reckoning is essentially a conceptualization of the social structure, the points of reference being a projection into the past of actual relations between social groups. 'It is less a means of co-ordinating events than of co-ordinating relationships, and is therefore mainly a looking-backwards, since relationships must be explained in terms of the past.'[15]

These and other examples reveal that just as our intuition of space is not unique, for we now know that there is no unique geometry that we must necessarily apply to space, so there is no unique intuition of time that is common to all mankind. Not only primitive people but relatively advanced civilizations too have assigned different degrees of significance to the temporal mode of existence and to the importance or otherwise of temporal perspective. In short, time in all its aspects has been regarded in many conceptually distinct ways.

2. Describing Time

Time, language, and number

Despite its corroboration of the view that ours is not the only way of regarding time, Whorf's study of the Hopi provides powerful evidence of a universal connection between time and language. Similarly, despite the great diversity of existing languages and dialect, the capacity for language appears to be identical in all races. Consequently, we can conclude that man's linguistic ability existed before racial diversification occurred.

In a famous paper on 'The Problem of Serial Order in Behaviour', delivered at the Hixon Symposium on Cerebral Mechanisms in Behavior in 1948, the American physiological psychologist K. S. Lashley argued that the organizing principle underlying the problems of syntax in speech and language is essentially rhythmic in nature, a view which is now generally accepted. Lashley's pioneer investigation of the temporal aspects of language has been developed further by the American physiologist E. H. Lenneberg, notably in his seminal book *Biological Foundations of Language*, published in 1968. Lenneberg has pointed out that many physiological processes which might be thought to have no temporal aspect do, in fact, exhibit one: for example, in the process of seeing, which appears to be instantaneous, time plays a role, for the identification of even the simplest shapes requires temporal integration in the nervous system. Like Lashley, he believes that the foundations of linguistics are to be found in our anatomy and physiology. In his words, 'Language is best regarded as a peculiar adaptation of a very universal physiological process to a species-specific ethological function: communication among members of our species.'[1] Lenneberg came to the conclusion that human articulation involves a basic periodicity of about six cycles a second (with a possible variation of up to a cycle from one individual to another), and he showed that a great variety of phenomena could be explained by this hypothesis.

In his Clayton Memorial Lecture on 'Some Aspects of Speech', which he delivered to the Manchester Literary and Philosophical Society in

1959, C. M. Bowra pointed out that the vocabularies of most primitive peoples are much more extensive than those used by modern sophisticated Europeans and that the reason for this is that, although they have no words for abstract concepts, they tend to be extremely subtle in their detection of fine distinctions in the visible world, which they denote by separate words. Their highly complex languages suit them very well so long as they are not obliged to come to terms with novel and unprecedented conditions. Since the state of equilibrium between survival and starvation which they normally experience is often finely balanced, it is not surprising that they usually consider it dangerous to deviate from their traditional customs and habits. Because they tend to adapt their lives and way of thinking to circumstances which they believe to be immutable, their rules and customs inevitably become rigid. Consequently, their languages, which are intimately adjusted to their way of life, tend to prevent the free movement of their minds into new regions of experience. As Bowra points out:

In so far as these languages change, and they certainly do, it is towards an ever greater elaboration in their own special methods of dealing with individual impressions and with the finer shades of difference in social relations. It is not surprising that men who spoke them were quite unable to understand what was happening when white men shot them for breaking rules which were to them totally unintelligible.[2]

It is now generally recognized that language is man's most outstanding characteristic. The possibility of human language seems to have depended not only on the potentialities of the vocal tract in man but also on the development of Broca's area in the neo-cortex. This area is thought to be concerned with the regulation of *sequences* of sounds. If this is so, the apparent lack of such an area in the brain of other primates may explain why the calls of these animals are not formed by varying the order in time of elementary units.[3]

Children are born with a general facility for language in so far as they exhibit an irresistible drive to express themselves. The babbling of infants is a spontaneous reflex activity, broadly similar to the uncoordinated movements of their limbs. It is an obvious, but none the less remarkable, fact that every normal child has the inherent ability to produce all the sounds of every language in the world, of which there are several thousands. Nevertheless, it is only the child's 'mother tongue' which he learns to speak spontaneously, and he must begin to do this before the age of about 6, as has been shown by the failure to learn to speak of the so-called 'wolf children' who have been unable to make contact with other human beings before that age. Every other language that one tries to learn later requires a special effort. Nevertheless language-learning comes easily only to some. The maximum number of languages that any one man is definitely known to have acquired is just under sixty. The famous orientalist Sir William Jones (1746–94) is said to have known over forty.

The reason why speech is based on sound, rather than gesture, is probably because sound is the sense most closely related to time. Nevertheless, although sound is transitory, the development of language originally depended on man's recognition of long-enduring objects to which names could be given, for there is good reason to believe that the introduction of verb-tenses was a comparatively late development. Our knowledge of the evolution of language is necessarily confined to written records, but they support this conclusion. For example, in Middle Egyptian of about 2000 BC, the 'tenses' were concerned with the repetition of the notion expressed by the verb rather than with the temporal relation of the action concerned to the time associated with the speaker. This was not just a peculiarity of Middle Egyptian, for we find that in other ancient forms of language the dominant temporal characteristic was duration rather than tense. Indeed, it is only in Indo-European languages that distinctions between past, present, and future have been fully developed. In Hebrew, for example, the verb treats action not in this way but as either incomplete or perfected. Moreover, 'the future is preponderantly thought to lie before us, while in Hebrew future events are always expressed as coming after us.'[4] On the other hand, already in archaic Greek we find evidence of verbal forms that discriminated between the tenses.

'Old English', the language spoken in England before the Norman Conquest, contained no distinct words for the future tense. Instead, the present tense was specially adapted for that purpose as and when necessary.

Suzanne Fleischman has drawn attention to the fact that the tenses we now use correspond to distinct mental activities: the past to knowledge; the present to feeling; and the future to desire and obligation, as well as potentiality. Owing to the stress laid by Christianity on moral obligation, it has been claimed that the rise of that religion was the sole reason that new modal futures were introduced about the fifth century AD, but in her opinion no less importance should be assigned to the effect of the shift that occurred about the same time in the basic word-order of Latin sentences from SOV (subject–object–verb) to SVO. She considers that 'an appeal to multiple causation—not ruling out the possibility of cultural determinants—may well prove to be the most satisfactory approach to the problem.'[5]

George Steiner has recalled the shock he experienced when, as a young child, he first realized that statements could be made about the *far* future. 'I remember', he writes, 'a moment by an open window when the thought that I was standing in an ordinary place and "now" and could say sentences about the weather and those trees fifty years on, filled me with a sense of physical awe. Future tenses, future subjunctives in particular, seemed to me possessed of a literal magic force.' He compares that feeling with the mental vertigo which is often produced by contemplating extremely large numbers, and draws attention to the interesting suggestion made by some scholars of Sanskrit, the oldest Indo-European language known, that 'the development of a grammatical system of futurity may have coincided with an interest in recursive series of very large numbers'.[6]

Be that as it may, it is clear that the origin of the concept of number, like the origin of language, is closely connected with the way in which our minds work in time, that is, by our being able to attend, strictly speaking, to only one thing at a time and our inability to do this for long without our minds wandering. Our idea of time is thus closely linked with the fact that our process of thinking consists of a linear sequence of discrete acts of attention. As a result, time is naturally associated by us with counting, which is the simplest of all rhythms. It is surely no accident that the words 'arithmetic' and 'rhythm' come from two Greek terms which are derived from a common root meaning 'to flow'. The relation between time and counting is further discussed in my *The Natural Philosophy of Time*.[7]

Time and natural bases of measurement

Most people, however primitive, have some method of time-recording and time-reckoning based either on the phases of nature indicated by temporal variations of climate and of plant and animal life or on celestial phenomena revealed by elementary astronomical observations. Time-reckoning, that is the continuous counting of time-units, was preceded by time-indications provided by particular occurrences. The oldest method of counting time was by means of some readily recognizable recurrent phenomenon, for example the counting of days in terms of dawns such as we find in Homer ('This is the twelfth dawn since I came to Ilion', *Iliad*, xxi. 80–1). In this method of time-reckoning, as M. P. Nilsson has remarked, it is not the units as a whole that are counted, since the unit as such has not been conceived, but a concrete phenomenon occurring only once within this unit. It is what he calls the '*pars pro toto* method', so extensively used in chronology.[8]

A good example of this method is provided by the extended use of the word 'day'. The fusion of day and night into a single unit of twenty-four hours did not occur to primitive man, who regarded them as essentially distinct phenomena. It is a curious fact that even now very few languages have a special word to denote this important unit. Notable exceptions are the Scandinavian terms, for example the Swedish *dygn*, whereas in English we use the same word 'day' to denote the full twenty-four-hour period and also the daylight part of it. Instead of appealing to 'dawn' and 'day', some peoples count time by the number of nights. This may be because sleeping provides a particularly convenient time-indicator. A familiar relic of this in English is the word 'fortnight', a term which is now as obsolete in the United States as the word 'sennight' is in Britain.

To indicate a particular time in the period of daylight the sun can often be used, either by reference to its position in the sky or in some other way. Thus, the Australian aborigine will fix the time for a proposed action by placing a stone in the fork of a tree so that the sun will strike it at the required time. Many tribes in the tropics indicate the time of day by referring to the direction of the sun or to the length or position of the shadow cast by an upright stick, but before sunrise the natural phenomenon most widely used as a time-indicator is cock-crow.

A wide variety of conventions have been adopted for deciding when the day-unit begins. Dawn was chosen by the ancient Egyptians, whereas sunset was chosen by the Babylonians, Jews, and Muslims. The Romans at first chose sunrise but later midnight, because of the variable length of the daylight period. Dawn was the beginning of the day-unit in Western Europe before the advent of the striking clock in the fourteenth century, but later midnight was chosen as the beginning of the civil day. Astronomers, such as Ptolemy, found it more convenient to choose midday, and this remained the beginning of the astronomical day until 1 January 1925 when, by international agreement, the astronomical day was made to coincide with the civil day.

Besides the day the other most important natural unit of time is the year. Nevertheless, although each year normally presents the same cycle of phenomena, man only gradually learned to unite the different seasons into a definite temporal unit. This step was particularly difficult to take by people living in those equatorial regions where there are two similar half-years, each with its own seed-time and harvest, since by a 'year' a vegetation-period was originally understood. There is an important difference between the natural year, that is, the period of the earth's annual revolution around the sun, and the agricultural year. The former has no natural beginning or end, whereas the latter has. In Old Norse, German, and Anglo-Saxon years tended to be reckoned in winters. The reason for this practice, which was of course rare in the tropics, was the same as that for counting days by nights, winter being a season of rest, an undivided whole, and therefore more convenient than summer with its many activities. Nevertheless, there were exceptions to this rule. For example, in Slavonic time was reckoned in summers and in English expressions such as 'a maiden of eighteen summers' were used, whereas in medieval Bavaria years were reckoned in autumns.

Time-indications from climatic and other natural phases during the course of the year are only approximate and tend to fluctuate from year to year. Greater accuracy is often desirable for agriculture, and it was recognized long ago that this could be provided by the stars, particularly by their rising and setting. Observation of these phenomena did not make great intellectual demands on primitive man, who rises and goes to bed with the sun. Experience teaches him which stars rise in the east just before the sun and which appear in the west at dusk and shortly afterwards set there. These 'heliacal' risings and settings, as they are called, vary throughout the year and can be readily correlated with particular natural phenomena. The stars therefore provide us with a ready and more accurate means of determining the time of year than any based on the phases of terrestrial phenomena. Just as the time of day may be revealed by the position of the sun, so the time of year can be determined by means of heliacal risings and settings, and this can form the basis of a calendar. Timings can also be approximately determined by observing the position of stellar groupings that can be easily recognized, notably the Pleiades.

Although the stars can help man to determine the seasons, they do not enable him to divide the year into parts. Instead, the moon has been used to produce a temporal unit between the year and the day. Moreover, unlike time-indications from natural phases and the stars, the moon's waxing and waning provide a continuous means of time-reckoning. Consequently, the moon can be regarded as the first chronometer, since its continually changing appearance drew attention to the durational aspect of time. Although the concept of the month is much more readily attained than that of the year, it is difficult to combine the two satisfactorily, because the solar period is not a convenient multiple of the lunar period. So long as the beginning of the month was determined by observing the new moon, the month ,was based on lunations, but they are inconvenient for measuring time, since it is the movement of the sun that determines the seasons and the rhythm of life associated with them. As a result, our system of months no longer has any connection with the moon but is a purely arbitrary way of dividing the solar year into twelve parts. Our present concept of the year can be traced back to the Romans and through them to the Egyptians, who disregarded lunation as a time-measure.

As regards shorter intervals of time than the year and the day, primitive people have often made use of convenient physiological intervals such as 'the twinkling of an eye' or occupational intervals such as the time required for cooking a given quantity of rice. Indeed, man's unwillingness to abandon natural bases of measurement was for long a hindrance to the development of a scientific system of timekeeping. This is particularly evident in the case of the hour. The division of the daylight period into twelve parts was introduced by the Egyptians, who first of all divided the interval from sunrise to sunset into ten hours and then added two more for morning and evening twilight respectively. They also divided the night into twelve equal parts. These 'seasonal hours', as they are called, varied in duration according to the time of year. The inconvenience of this practice, although not so great in countries like Egypt as in more northerly places, introduced an unnecessary complication into the development of the water-clock and was quite impracticable in scientific astronomy.

Time in contemporary society

What particularly distinguishes man in contemporary society from his forebears is that he has become increasingly time-conscious. The moment we rouse ourselves from sleep we usually wonder what time it is. During our daily routine we are continually concerned about time and are forever consulting our clocks and watches. In previous ages most people worked hard but worried less about time than we do. Until the rise of modern industrial civilization people's lives were far less consciously dominated by time than they have been since. The development and continual improvement of the mechanical clock and, more recently, of portable watches has had a profound influence on the way we live. Nowadays we are governed by time-schedules and many of us carry diaries, not to record what we have done but to make sure that we are at the right place at the right time. There is an ever-growing need for us to adhere to given routines, so that the complex operations of our society can function smoothly and effectively. We even tend to eat not when we feel hungry but when the clock indicates that it is meal-time. Consequently, although there are differences between the objective order of physical time and the individual time of personal experience, we are compelled more and more to relate our personal 'now' to the time-scale determined by the clock and the calendar. Similarly, in our study of the natural world, never has more importance been attached to the temporal aspects of phenomena than today. To understand why this is so and how it has come about that the concept of time now dominates our understanding of both the physical universe and human society, no less than it controls the way we organize our lives and social activities, we must examine the role that it has played throughout history.

Part II
Time in Antiquity and the Middle Ages

3. Time at the Dawn of History

Prehistory

Consciousness of self is a fundamental characteristic of human existence. It involves a sense of personal continuity through a succession of different states of awareness. This sense of personal identity depends essentially on memory, but a sense of the past could only have arisen when man consciously reflected on his memories. Similarly, purposeful action involves at least implicit recognition of some future achievement, but a general sense of the future could not have resulted until man applied his mind systematically to the problem of future events. Man must have been conscious of memories and purposes long before he made any explicit distinction between past, present, and future.

The famous palaeolithic paintings found in caves such as that at Lascaux in the Dordogne have been interpreted as evidence that, at least implicitly, people were operating 20,000 or more years ago with teleological intent in terms of past, present, and future. From what we know of primitive races it is highly probable that the incentive for producing these paintings was magical, the object being to fix in paint on the wall or ceiling of a cave an event—usually the slaying of an animal—which it was hoped would be effected in the future elsewhere. It may be that those responsible for the well-known picture of the so-called 'Dancing Sorcerer' (on the wall of one of the innermost recesses of the Trois Frères cave in the department of Arriège in France), which represents a man in the skin of an animal and wearing the antlers of a stag, may have felt that the actual performance of the dance was insufficient, since they were concerned about the conservation of the magical efficacy of the dance after it had ended. If correct, this hypothesis might explain why these people so many thousands of years ago went to the trouble and danger of penetrating so deeply into the cave for this purpose.

In making these pictorial representations people must have relied on their memories of past events, and so all three modes of time were involved. But this no more implies a conscious awareness of the distinctions between past, present, and future than the use of language necessitates an explicit knowledge of grammar. Indeed, it must have required an enormous effort for man to overcome his natural tendency to live like the animals in a continual present. Moreover, the development of rational thought actually seems to have impeded man's appreciation of the significance of time.

In his classic work *Primitive Man as Philosopher*, Paul Radin argues that among primitive men there exist two different types of temperament: the man of action who is oriented towards external objects, interested primarily in practical results and comparatively indifferent to the stirrings of his inner self, and the thinker—a much rarer type—who is impelled to analyse and 'explain' his subjective states. The former, in so far as he considers explanations at all, inclines to those that stress the purely mechanical relations between events. His mental rhythm is characterized by a demand for endless repetiton of the same event or events, and change for him means essentially some abrupt trans-formation. The thinker, on the other hand, finds purely mechanical explanations inadequate. But, although he seeks a description in terms of a gradual development from one to many, simple to complex, cause to effect, he is perplexed by the continually shifting forms of external objects. Before he can deal with them systematically he must give them some permanence of form. In other words, the world must be made static.[1]

Belief that ultimate reality is timeless is deeply rooted in human thinking, and the origin of rational investigation of the world was the search for the permanent factors that lie behind the ever-changing pattern of events. As Radin stressed in his discussion of the thought of primitive man, 'as soon as an object is regarded as a dynamic entity, then analysis and definition become both difficult and unsatisfactory. Thinking is under such circumstances well-nigh impossible for most people.'[2] Indeed, language itself inevitably introduced an element of permanence into a vanishing world. For, although speech itself is transitory, the conventionalized sound symbols of language transcended time. At the level of oral language, however, permanence depended solely on memory. To obtain a greater degree of permanence the time symbols of oral speech had to be converted into the space symbols of written speech. The earliest written records were simply pictorial representations of natural objects, such as birds and animals. The next step was the ideograph by means of which thoughts were represented symbolically by pictures of visual objects. The crucial stage in the evolution of writing occurred when ideographs became phonograms, that is representations of things that are heard. This conversion of sound symbols in time to visual symbols in space was the greatest single step in the quest for permanence.

The distinctions we make between past, present, and future refer to the transitional nature of time. Although dependent on memory, our sense of personal identity is closely associated with the durational aspect of time. Man's discovery that he himself, like other living creatures, is born and dies must have led him intuitively to try to circumvent the relentless flux of time by seeking to perpetuate his own existence indefinitely. Evidence of ritual burial goes back at least to Neanderthal man and possibly even earlier.[3] A Neanderthal burial of about 60,000 years ago, at a cave in northern Iraq, even appears to have included flowers.[4] As for our own species, the oldest evidence, going back to possibly about 35,000 BC, reveals that the dead were not only equipped with weapons, tools, and ornaments but also with food, which must often have been in short supply among the living. In some cases bodies were covered with red ochre, which may have been intended to simulate blood, in the hope of averting physical extinction. The care taken over the disposal of the dead indicates a deeply held conviction that, provided the appropriate steps were taken, death could be regarded as a transitional state.

The idea of death as a transition from one phase of life to another that could only be satisfactorily effected by performing the appropriate rituals became the pattern for dealing with other natural changes. The principal transitions from one phase of people's life to another were thought of as crises and as a result the community to which they belonged assisted with the appropriate rituals.

Similarly, the principal transitions in nature were also regarded as occurring suddenly and dramatically. In the palaeolithic period men were already aware that at certain times of the year animals and plants are less prolific than at others, and seasonal ritual observances to maintain an adequate supply of them were therefore deemed necessary. With the change from a nomadic and food-gathering to an agricultural and more highly organized form of society, man's anxiety about himself and the animals that he hunted merged into a wider anxiety about nature. At the critical seasons a ritual response was required to overcome the unpredictable factors that might otherwise interfere with the regular growth of crops. The succession of natural phenomena and phases became evidence for a dramatic interpretation of the universe. Nature was seen as a process of strife between divine cosmic powers and demoniacal chaotic powers in which humans were not just spectators but were obliged to play an active part in helping to bring about the required phenomena by acting in full unison with nature. This meant performing a given set of rituals at the appropriate times.

In recent years the study of megalithic remains such as Stonehenge in terms of hypothetical astronomical alignments has led to various interesting speculations concerning prehistoric man's knowledge of the calendar. A careful assessment of these views has been made by D.C. Heggie.[5] It has even been suggested that many of the markings found on upper palaeolithic artefacts and in caves are probably calendrical or astronomical in nature.

Ancient Egypt

In the oldest civilizations we find definite correlations between social and natural events. In Egypt, where everything depended on the Nile, the coronation of a new pharaoh was often postponed until a new beginning in the cycle of nature provided a propitious starting-point for his reign. It was made to coincide either with the rising of the river in early summer or with the recession of the waters in autumn when the fertilized fields were ready to be sown. The royal ritual was closely associated with the history of Osiris, the divine prototype on whom the pharaohs modelled themselves by re-enacting his traditional deeds. Osiris represented the life-giving waters and the soil fertilized by the Nile. After the Nile had receded the land eventually appeared to die, but on the reappearance of the waters it revived again. The Osiris myth embodied this cycle of birth, death, and rebirth and gave the promise of immortality. At death a series of rites enabled the pharaoh himself to become Osiris and thereby safe from the depredations of time. At first this way to immortality was essentially a royal prerogative, but eventually similar rites were thought to confer immortality on anyone who could afford to imitate them. As S. G. F. Brandon has pointed out, the great popularity of the cult of Osiris meant, in effect, the adoption by the Egyptians of a definite concept of time, although this may not have been consciously recognized. For, since the Egyptians believed that Osiris had actually lived in their land long ago, his cult signified that a particular historical event, in this case the death and resurrection of Osiris, could be perpetually repeated by magical simulation so that its supposed good effects could benefit those persons on whose behalf the rites were performed.[6]

Although the Osiris cult was a striking instance of what Brandon called the 'ritual perpetuation of the past', it was concerned only with personal immortality and generated no interest in the past as such. On the contrary, by trying to re-create on specific occasions particular events associated with Osiris thought was concentrated on the present rather than on the past. The Egyptians regarded time as a succession of recurring phases. They had very little sense of history or even of past and future. For, although there was an absolute past, it was normative and was not regarded as receding.[7] They thought of the world as essentially static and unchanging. In the beginning the gods created the world with everything in it organized on a permanent pattern. The cosmic balance, which involved the regular recurrence of the seasonal phenomena, could, however, only be maintained by an unceasing control. On earth this was the function of the pharaoh. Historical incidents were no more than superficial disturbances of the established order or recurring events of unchanging significance. This idea of a perpetually repetitive pattern of events inspired a sense of security from the menace of change and decay. If some crisis occurred to disturb the customary order of things, it could not be something really new but was foreseen at the creation of the world. The priests would therefore examine ancient writings to find out if the event had already occurred in the past and what solution had then been applied to it.

This evaluation of the Egyptians' attitude to time is borne out by their attitude to chronology. The years were not numbered in a linear succession but according to a particular pharaoh's reign, each mounting the throne in the year 1, and also according to the levy of taxes. The treasury officials numbered the royal possessions every two years, so that the years of a given reign were designated as the Year of, say, the Third Numbering, or the Year after the Third Numbering, and so on. This absence of a continuous sense of time made an exact computation of past centuries extremely difficult, particularly because of co-regencies, parallel reigns and fictitious reigns. When they said, for example, 'in the reign of the King Cheops' they thought of a distant event situated in time in a rather vague way. Furthermore, the idea the Egyptians had of an eternal and immutable world meant that they never imagined any evolution of social conditions. There were periods of considerable social disturbance, particularly at the end of the Old Kingdom, but only the literary texts mention them. The historical texts were confined to enumerating the kings who lived in those troubled years and do not give any indication that something important was occurring at that time. For nearly 3,000 years the recording of historical events by the Egyptians was characterized by a preoccupation with royal lists and a lack of precise dates. Only one Egyptian historian is known to us, the priestly scribe Manetho who compiled the list of all the pharaohs and conveniently divided them into the particular groups or dynasties which Egyptologists still employ today. But Manetho, who lived in the third century BC, wrote in Greek and his work must be regarded as Hellenistic in character rather than Egyptian.

Nevertheless, in one respect the Egyptians made an outstanding contribution to the science of time. For they devised what Otto Neugebauer has described as 'the only intelligent calendar which ever existed in human history'.[8] Their civil year consisted of twelve months, each of thirty days, with five additional days at the end of each year, making 365 in all. In Neugebauer's view, it originated on purely practical grounds by continual observation and averaging of the time intervals between successive arrivals of the Nile flood at Heliopolis, the rising of the Nile being the main event in Egyptian life. At first the Egyptians did not realize that the astronomical year does not consist of exactly 365 days but contains an extra fraction (about one-quarter) of a day. The discrepancy was soon recognized and another calendar was then introduced which kept more closely in phase with astronomical phenomena. It was noted that the rising of the Nile occurred when the last star to appear on the horizon, before dawn obscures all stars, was the dog star Sothis, or Sirius as it is known to us. This 'heliacal rising', to use the term employed in Greek astronomy, thus came to be regarded as the natural fixed point of the 'Sothic' calendar. Astronomical computations show that the first day of the two calendars agreed in the year 2773 BC, and it has been concluded that this was when the Sothic calendar was introduced.[9] There is reason to associate this with the Minister of King Zoser of the Third Dynasty known as Imhotep, later deified as the Father of Egyptian science. The Sothic calendar kept pace with the seasons, but the civil calendar did not. The two coincided at intervals of 1460 (= 365 × 4) years. The civil year was divided into three conventional 'seasons'—called time of inundation, sowing time, and harvest time—and each of them was divided into four months, these being of course conventional too and not connected with the moon. Despite the linguistic anomaly that the season called 'the time of inundation' would in due course fall in one of the other seasons, the Egyptians retained the 365-day calendar right down to the Roman period because of its convenience as an automatic record of the passage of time in an era, each year containing the same number of days, unlike our years. This calendar was just what was needed for astronomical calculations. It was taken up by the Hellenistic astronomers, became the standard astronomical system of reference in the Middle Ages, and was even used by Copernicus in his lunar and planetary tables. The Egyptians also had a lunar calendar to regulate festivals by phases of the moon. They found that 309 lunar months were almost equal to twenty-five civil years.

In a nearly cloudless country such as Egypt observation of the sun was a useful way of telling the time and it is therefore not surprising that the earliest known solar clock has been found there. A fragment of an Egyptian sun-clock dating from about 1500 BC is now in a museum in Berlin. Shaped like a T-square, it was placed horizontally with the crosshead laid towards the east in the forenoon, thereby casting a shadow along the stem which was graduated with marks for six hours. As the sun rose higher in the sky the shadow shortened until noon, when it disappeared at the sixth hour mark. Then the instrument was relaid with the crosshead towards the west so that the lengthening shadow gradually moved back along the hour marks to the twelfth. The earliest clocks of this type were correct only at the equinoxes, and not until much later was it possible to take due account of the seasonal changes in the position of the sun. Eventually a series of hour scales, seven in number, was devised to accommodate these changes, but even then this timepiece was seldom accurate. The warrior pharaoh Tuthmosis III referred to the hour indicated by the sun's shadow at a critical juncture of one of his campaigns in Asia, and it would therefore seem that he carried with him a portable sun-clock.[10] Another form of sun-clock employing the direction rather than the length of the sun's shadow was the sundial, but the Egyptians who invented it were far from understanding the subtleties involved in making an accurate instrument of this type, which must be calibrated for the latitudes of the different places where it is to be used.

To provide a means of measuring time at night the Egyptians also invented the water-clock, or 'clepsydra' as the Greeks later called it. Two main types were developed, depending on whether water flowed out of or into a graduated vessel. Whereas inflow clocks were usually cylindrical, outflow clocks were in the form of inverted cones with a small hole at or near the bottom, the time being indicated by the level of water. Clepsydrae were also used by the Greeks and Romans. Vitruvius, writing about 30 BC, described a number of types. To make them indicate seasonal hours, either the rate of flow or the scale of hours had to be varied according to the time of year, and considerable ingenuity appears to have been applied to achieve this.

The Egyptians also used a plumb-line, which they called the 'Merkhet', to determine the time at night. They observed the transits of selected stars across the meridian as they came into line with two Merkhets. A Merkhet is on permanent exhibition in the Science Museum, London. It is thought to date from about 600 BC. According to the inscription that it bears, it belonged to the son of a priest of the Temple of Horus at Edfu, in Upper Egypt.

As mentioned in chapter 2, we are indebted to the Egyptians for our present division of the day into twenty-four hours, although the Egyptian hours were not of equal length, since at all times of the year the periods of daylight and darkness were each divided into twelve hours. The end of the night was marked by the heliacal rising of a particular star. However, because the sun not only participates in the daily rotation of the heavens from east to west but also has its own slow annual motion relative to the stars in the opposite direction, different heliacal risings occur throughout the year. Instead of choosing a different star daily, the Egyptian priests, who were primarily concerned with the timing of the nightly service in their temples, made a fresh choice every ten days, a period of time (and stellar constellation) known as a 'decan'. The oldest astronomical texts now known are found on the lids of wooden coffins dating from the Ninth Dynasty (c.2150 BC). They are called 'diagonal star-clocks', or 'diagonal calendars', and they give the names of the stars associated with the respective decans. These star charts were provided to enable the deceased to tell the time of night or the date in the calendar.[11] Incidentally, the twelve signs of the zodiac did not appear in Egypt until the Hellenistic period, nor is there any trace of astrological ideas there before then.

Since the Egyptian civil year contained 365 days, there were thirty-six decans in the year (plus the five extra days at the end of the year), and the sky was divided accordingly. During the summer, when Sirius rises heliacally, only twelve of these divisions of the sky can be seen rising during the hours of darkness, and it was this that led to the twelve-hour division of the night. As for the period of daylight, a simple sundial on an obelisk of Seti I, about 1300 BC, indicates ten hours between sunrise and sunset, to which two more were added for morning and evening twilight. As previously mentioned, these divisions of the day and night led to the twenty-four 'seasonal' hours of the complete day in Hellenistic and Roman times. In antiquity only the Hellenistic astronomers used hours of equal length, these being the same as the seasonal hours at the date of the spring equinox. Since, following Babylonian practice, all astronomical computations involving fractions were conducted in the sexagesimal system, instead of our current decimal system, these 'equinoctial' hours were divided by the astronomers into sixty firsts, or minutes, and each of these was subdivided into sixty seconds. Thus, as Neugebauer has succinctly remarked, our present way of dividing up the day into hours, minutes, and seconds 'is the result of a Hellenistic modification of an Egyptian practice combined with Babylonian numerical procedures'.[12]

Sumeria and Babylonia

Although there was always the possibility of drought or flooding, the Nile seldom brought disaster to Egypt. Mesopotamian civilization developed in a very different environment. The Tigris and Euphrates are far less uniform in their behaviour than the Nile. The inhabitants of ancient Mesopotamia had to contend with variations of climate, scorching winds, torrential rains, and devastating floods over which they had little control. The mood of Mesopotamian civilization reflected this element of force and violence in nature which gave no grounds for believing that the ravages of time could be surmounted by a ritual cult like that of Osiris in Egypt. Although there was evidence of cosmic order in the motions of sun, moon, and stars and in the cycle of the seasons, this order was not regarded as securely established but had continually to be achieved by the integration of conflicting divine wills or powers.[13] The basic framework of society in Mesopotamia remained the same for 2,000 years or more, but at different times Sumerians, Babylonians, and Assyrians were dominant and the order of society was far less static than in Egypt. Whereas in Egypt the pharaoh symbolized the triumph of an invincible divine order over the forces of chaos, in Mesopotamia kingship represented the struggle of a human order with all its anxieties and hazards to integrate itself with the universe.[14]

The sense of insecurity which affected the city-states of Mesopotamia led to a rudimentary interest in the history of social order. This is revealed in texts going back to about 2000 BC, notably in the 'Sumerian King List' which begins with a sequence of eight kings, presumably fabulous, whose reigns add up to a total of 241,200 years![15] The sequence was then interrupted by a flood that was so devastating that a new start had to be made and again kingship had to be 'lowered from heaven'.

Archaeological evidence has revealed that a cataclysmic flood over-whelmed the Sumerian plains about 4200 BC.

Despite their interest in this past event and the compilation of chrono-logical lists of kings with grandiloquent accounts of their achievements, the Sumerians and their successors were not really historically minded. They were mainly interested in themselves and were content to leave historical matters relatively indefinite.[16] Their reason for perpetuating the memory of the Flood was most probably magical. A destructive flood was an annual possibility, and the god of Heaven Anu and the storm-god Enlil, who were believed to have been responsible for the decision to destroy mankind, were invoked in the incantation passages of the legend of the Flood. Similarly, although libraries were established in temples and palaces in order to conserve records of the past, there is no evidence of any interest in history, except in so far as it was a guide to action in the present. Indeed, the general conception of the cosmic process envisaged by the ancient inhabitants of Mesopotamia precluded the possibility of history having any ultimate significance or purpose. The apparent lack of any meaning in its repetitive pattern is expressed in the following passage from the *Epic of Gilgamesh*: 'There is no permanence. Do we build a house to stand for ever, do we seal a contract to hold for all time? Do brothers divide an inheritance to keep for ever, does the flood-time of rivers endure? . . . From the days of old there is no permanence.'[17]

Although in Mesopotamia kingship was never so important as in Egypt, its function was the maintenance of harmony between earth and heaven. There were at one time a number of city-states, each with its own god. Unified rule was ultimately achieved by Hammurabi towards the beginning of the second millenium BC, with its centre in Babylon. In cosmic terms this implied the ascendancy of Marduk, the god of Babylon, over the other gods. As a result, the most important ritual in Mesopo-tamia was the spring New Year Festival at which the epic of the creation of the world by Marduk was recited. The significance of this epic was not as a record of the past, but rather as a means of ensuring the theologico-political supremacy of Marduk in the present. For Marduk was not the most ancient of the gods, and his lordship over the other gods was meant to justify the political supremacy that Babylon had acquired.

Although the New Year Festival symbolized the inauguration of a new solar cycle, the renewal of fertility, and victory over chaos, its celebration provided no guarantee that the social order would continue undisturbed. The king and his counsellors therefore watched for portents that could be interpreted, so that disasters might be foreseen and, if possible, averted. There was assumed to be a counterpart in human events to every celestial phenomenon. This belief led the priests to make careful and systematic observations of the heavenly bodies. Celestial omens began to be used as portents on a considerable scale in the first Babylonian dynasty (eighteenth to fifteenth centuries BC), although lunar eclipses may have been regarded as ominous previously.[18] The prediction of this so-called 'judicial' astrology referred to the royal court and the state and not to ordinary individuals. Horoscopic astrology, according to which the positions of the planets at the time of birth determines the fate of the individual, did not develop until much later. The oldest known horoscope goes back to 410 BC when Babylonia was part of the Persian empire.[19] In Hellenistic and Roman times the Chaldeans, as the Babylonians were called, came to be regarded as the great experts in astrology. Both the older judicial astrology and the later horoscopic astrology were based on a fundamentally deterministic, or fatalistic, view of existence. People who believe that history and the destinies of men are controlled by the stars are not likely to entertain the idea of historical progress. Instead, they are more inclined to adopt a cyclical view of time, in accordance with the periodicity of the motions of the sun, moon, and planets. To what degree, however, such a view of time was developed in Mesopotamian thought is not revealed by the cuneiform records, although according to Seneca the late Babylonian astronomer-priest Berossus (*c*.300 BC) believed in the periodic destruction and re-creation of the universe.[20]

The heavens were studied not only for omens but also for the sake of the calendar. The basis of the Babylonian calendar seems always to have been lunar. The month began when the new lunar crescent was for the first time visible again after sunset. Consequently, the Babylonian day began in the evening. A lunar month defined in this way must contain a whole number of days, but sometimes this was twenty-nine and sometimes thirty. To solve this problem the motion of the sun had to be investigated. The late Babylonian astronomers of the fourth and later centuries BC studied the motions of the sun and planets with great care and mathematical ingenuity, but their most detailed investigations were of the moon because the calendar was moon-based. They invented harmonic analysis, in the sense that they introduced the idea of breaking down a complicated periodical effect into a sum of simpler periodical effects in order to make the mathematics tractable. They did not use trigonometrical methods but linear 'zigzag functions'.[21]

The lunar 'year' normally comprised twelve months, but this is less than the solar year. In order to prevent the seasons getting out of phase, a thirteenth month was inserted from time to time, but there was no regular system for the intercalation of this additional month until the fifth century BC, when seven of these months began to be inserted at fixed intervals in a cycle of nineteen years. Previously it is probable that the state of the harvest decided the need for the additional month. The nineteen-year cycle depends on the discovery that nineteen solar years are very nearly equal to 235 lunar months. It is usually known as the Metonic cycle, after the Athenian astronomer Meton who introduced it in 432 BC (see Appendix 2). Whether the cycle was discovered first by the Babylonian astronomer-priests or independently by them and Meton is uncertain.[22] The use of such a cycle by the late Babylonians shows that a fairly precise astronomical definition of a year was adopted by them. This was probably based on careful observation of the summer solstice. The invention of the zodiac, the belt round the sky in which the sun, moon, and planets lie, also occurred about this time.[23] The twelve zodiacal signs, of equal lengths of thirty parts each, are known to have been in use from soon after 500 BC. This division of the sky was eventually carried over to the division of the circle and so led to our present habit of dividing the complete (two-dimensional) angle around a point into 360 degrees.

The nineteen-year luni-solar cycle became the foundation of the Jewish and Christian calendars, since it solved the problem of establishing the dates of new moons for religious purposes. In particular, the origin of the problem of the dating of Easter can be traced back to the Babylonians. The rituals performed by the king-priest, particularly at the New Year Festival, were regarded as the repetitions of divine actions and were meant to correspond exactly in time as well as in character with the rituals on high. From this primitive idea sprang the belief that it was important to celebrate Easter at the correct date, since this was the crucial time of combat between God (or Christ) and the Devil, and God required the support of his worshippers to defeat the Devil.

The Babylonians paid particular attention to the seven-day periods associated with successive phases of the moon, each of these periods ending with an 'evil day' on which specific taboos were enforced so that the gods could be propitiated and conciliated. These prohibitive regulations were similar to those that many other peoples in different parts of the world have observed at changes in the appearance of the moon, but the Babylonians influenced the Jews, who in their turn influenced the early Christians and eventually ourselves. The ultimate origin of our seven-day week and the restrictions for long imposed on Sunday activities can thus be traced back to the Babylonians.

Ancient Iran

From 539 to 331 BC Babylonia was a part of the Persian empire. It was during this period that horoscopic astrology was invented, probably in the fifth century BC. For the casting of horoscopes one needs to know the positions of the planets for a given date. Often a horoscope is required for a date for which no observations are available, and horoscopic astrology therefore needs methods for computing the positions of the planets. The oldest known system of Babylonian planetary theory is thought to have been invented not earlier than 500 BC. It is possible that the motive was astrological and that the invention of horoscopic astrology at about this time was due to the influence of Iranian doctrines of the immortality and celestial origin of the soul.

The Iranians who conquered Babylonia were a branch of the Aryan race. Their native land consisted of a central plain surrounded by mountains. This plain was largely desert and was subject to extremes of climate. It was in this harsh and inhospitable land that one of the great religions of mankind originated. This religion, known as Zoroastrianism, involved a teleological interpretation of time. The date of its founder, Zarathustra (Zoroaster is the Greek form of his name), is uncertain but it is thought that he flourished in the first half of the sixth century BC. The Iranians already had a considerable religious heritage and it is difficult to decide how much of Zoroastrianism is due to the reforms introduced by Zarathustra.

Zarathustra belonged to a pastoral tribe in northern Persia. As a young man he had a prophetic revelation that led him to preach a new faith in place of the prevailing polytheism. He denounced the old religion as the Lie and called on men to worship the deity whom he called Ahura Mazdah, the wise Lord, who stood for the Truth. Zarathustra's monotheistic religion can be regarded as a response to the social conditions of his time, an age of transition when a settled agricultural and pastoral community was being threatened by predatory tribes who still followed the nomadic way of life. Zarathustra interpreted the struggle between good and evil forces in ethical terms, and he believed that it pervaded the whole universe. Although evil could not be attributed to Ahura Mazdah, its existence had to be accounted for, and Zarathustra explained it in terms of free will. At the beginning of time two spirits were created by Ahura Mazdah, the good spirit Spenista Mainyu (later called Ohrmazd) and the evil and destructive spirit Angra Mainyu (later called Ahriman). The latter, although his existence was due to God, became evil by his own free choice.

Zarathustra believed that man was involved in this cosmic struggle of good and evil and that he was compelled to choose one side or the other through his own conduct. This meant that man has an inescapable moral responsibility for his own actions. Zarathustra declared that at death God passes judgement on man and that this decides his fate when the world is finally transformed into the same state of perfection as when it left the hands of the Creator. At the last, immortal glory will be the reward of those who adhere to the Truth, whereas the followers of the Lie will be condemned to 'a long age of darkness, foul food and cries of woe'.[24] This doctrine of 'last things' was the first systematized eschatology in the history of religion and it profoundly influenced Judaism, Christianity, and Islam.

After Zarathustra's death his religion was taken up by the old priestly class known as the Magi and eventually became the faith of the Achaemenid dynasty. The first Persian king who seems to have accepted its basic doctrines was Darius (522–485 BC), but Achaemenid Zoroastrianism departed in certain respects from the original teachings of Zarathustra. There was some reversion to polytheism and the religion became more magical and ritualistic than ethical. Following the overthrow of the Achaemenid dynasty by Alexander of Macedon in 331 BC there was a confused period in the history of Zoroastrianism until its revival as the state religion under the Sassanian dynasty (AD 226–651). Most of the extant documents relate to this last period, which ended with the conquest of Persia by the forces of Islam.

Long before this happened there was a tendency to identify Ahura Mazdah with the good spirit Ohrmazd. This development gave rise to a perplexing problem. For Zarathustra had spoken of the good and evil spirits as twins and thereby implied that they had a common origin. A solution of this problem led to an important heresy associated with the idea of time, personified by the ancient god Zurvan. The significance of time was, of course, implicit in the eschatological character of Zoroastrianism. In the Zurvanite heresy it became the supreme deity. The reasoning that led to this development was clearly expressed in a remarkable passage in a late writing known as the Persian Rivayat:

Except Time all other things are created. Time is the creator; and Time has no limit, neither top nor bottom. It has always been and shall be for evermore. No sensible person will say whence Time has come. In spite of all the grandeur that surrounded it, there was no one to call it creator; for it had not brought forth creation. Then it created fire and water; and when it had brought them together, Ohrmazd came into existence, and simultaneously Time became Creator and Lord with regard to the creation it had brought forth.[25]

Throughout Iranian thought there was a tendency to dualism, and it is therefore not surprising that two distinct forms or aspects of time were recognized: indivisible time, that is the eternal 'now', and time that is divisible into successive parts. The former represented the creative aspect of time and was fundamental. It was called Zurvan akarana, or infinite time, and was the progenitor of the universe and of the spirits of good and evil. Associated with the universe was the other form of time called Zurvan daregho-chvadhata, that is time of the long dominion, or finite time. This was the time that brought decay and death. It dominated the world of man and was represented by the celestial firmament. Presumably under Babylonian influence, the life-span of time of the long dominion was set at 12,000 years, the number twelve corresponding to the twelve signs of the zodiac. This cosmic 'year' was divided into four periods, each of 3,000 years, Zarathustra's life occurring at the beginning of the final period.

The whole reason for the existence of finite time appears to have been to bring about that conflict of good and evil which eventually leads to the triumph of the former. A question that puzzled some followers of Zarathustra in later times was that, if Ohrmazd was all-powerful and so destined to overthrow Ahriman, why did this not happen immediately, so that the world would have been spared all the suffering caused by the conflict between them? An attempt to answer this question, and one which in the opinion of S. G. F. Brandon shows some consciousness of the significance of the factor of time, was made by a late Zoroastrian priest renowned for his orthodoxy. He argued that Ohrmazd, because his nature was good and just, could not destroy Ahriman until the latter had, by his evil deeds, provided just cause for his destruction.[26]

Finite time begins and ends with the rule of Ohrmazd. At a given moment finite time came into existence out of infinite time. It goes through a cycle of changes until it finally returns to its original state and then merges into infinite time. There is no evidence for any repetition of the cycle.

In later Zoroastrianism the emphasis laid by Zarathustra on the role of individuals and the character of their life was replaced by concern for the general fate of mankind. The unfolding of the divine purpose was not, however, identified with the course of human history as known to the inhabitants of Iran. Indeed Zoroastrians never attempted to relate the history of their nation to the cosmic struggle of Ohrmazd and Ahriman.

A problem concerning time and the calendar that has attracted much attention in recent years is the precise date when the Iranians adopted the Egyptian 'vague', or civil, calendar of 365 days. The Achaemenid monarch Cambyses conquered Egypt in the year 525 BC (his predecessor Cyrus had conquered Babylon in 539 BC) and so the replacement of the 'Old-Avestan calendar', as the one formerly used by the Iranians is now called, presumably occurred after that. The new calendar, now known as the 'Young-Avestan calendar' appears to have been adopted in the reign of Cambyses' great successor Darius I. The most thorough and plausible investigation of the date of its introduction is that made a few years ago by the distinguished historian of ancient astronomy the late Willy Hartner of Frankfurt University.[27] He came to the conclusion that the Young-Avestan calendar was introduced on 21 March 503 BC. (21 March is the 'Gregorian' date; the corresponding 'Julian' date would be 27 March.) In this calendar the year consisted of twelve months, each of thirty days except the eighth month, which comprised thirty-five days. The most important point, however, that emerged from Hartner's investigation was that by 503 BC the Babylonian astronomer-priests had discovered that the tropical year (the year of the seasons) is not of exactly the same length as the sidereal year (the 'true' astronomical year). This was an essential step towards the determination by the Hellenistic astronomer Hipparchus (*c.*150 BC) of the precession of the equinoxes, with its eventual important implications for the reform of the calendar in AD 1582 (see ch. 8).

According to the Iranian scholar S. H. Taqizadah, a correction was made to the Young-Avestan calendar in 441 BC to link it more closely with the seasons. It took the form of an intercalation of a full month of thirty days once every 120 years.[28] The Zoroastrians still reckon dates by the years of the last Zoroastrian king of Iran, Yazdgard III of the Sassanian dynasty (who was assassinated in AD 651 after the Arabs had overrun his kingdom), thus prolonging his imaginary reign through the centuries. In this way we know that the Persian New Year in AD 632, the year in which he ascended the throne, fell on 16 June.[29] The era and the Zoroastrian calendar are followed to this day by the Parsees of Bombay.

4. Time in Classical Antiquity

Classical Greece and Hellenistic civilization

About 1200 BC the late Bronze Age civilization of Mycenae, which had dominated the Aegean world since the destruction of Knossos some 300 years before, itself collapsed under the invasion of the Dorian Greeks from the north. The early Iron Age that followed lasted until about 800 BC, when the first city-states emerged. It was a dark age similar to the Dark Age of Western Europe following the final collapse of the Roman Empire. The Mycenaean past remained a folk-memory of the Greeks that was preserved orally, culminating in the epics of Homer. In them the one certain thing about man is his mortality, and this temporal restriction was the decisive factor that distinguished men from gods. Since the Greeks looked back on the Mycenaean past as a 'Golden Age' of gods and heroes, they tended to regard history as a decline from this ideal state and not as an ultimate order of reality.[1]

Consequently, for the Greeks, unlike the Persians, time was not a god. It only became a god in Hellenistic times when it was worshipped under the name 'Aion', but that signified a sacred, eternal time which was very different from ordinary time, *chronos*. Different thinkers had different ideas concerning the nature and significance of the temporal mode of existence. At the dawn of Greek literature two contrasting points of view are found in Homer and Hesiod.

In the *Iliad* Olympian theology and morality are dominated by space-like rather than time-like concepts, the cardinal sin being *hubris*, that is, going beyond one's assigned province. The whole conception, in the words of Cornford, is 'static and geometrical; everything has its limited field with bounds that must not be passed'.[2] Homer was not interested in the origin of things and had no cosmogony beyond the idea that water is the origin of all things. This was expressed mythologically by calling Oceanus, the river which encircles the world's disk, the origin of all things (*Iliad*, xiv. 246). W. K. C. Guthrie, who has drawn attention to this, makes the interesting point that it was probably an Ionian idea, since 'it reappears in Ionian philosophy and in the eastern peoples to whose influence early Ionia lay particularly open'.[3] It will be recalled that Thales, the first Greek philosopher, was Ionian, and he maintained that the first principle (*arche*) of all things is water.

Hesiod, unlike Homer, in his *Works and Days* gave an account of man's decline from a primeval Golden Age; his poem was based implicitly on the concept of time, although the word 'time' never actually appears in it. The main purpose of the poem was to offer advice concerning the regulation of the activities of the year, particular days being of good or evil omen, appropriate or inappropriate for different activities. In short, time was regarded by Hesiod as an aspect of the moral ordering of the universe.

Two centuries or more later, in the sixth century BC, the first Greek philosophers speculated, without invoking mythology, on how the world was generated. They regarded the world as being based on a single live space-filling substance from which all things developed spontaneously by the interplay of opposed processes such as separation and combination or rarefaction and condensation. The first explicit statement in Greek literature that, although individual things are subject to change and decay, the world itself is eternal appears to have been made by the philosopher Heraclitus about 500 BC. He regarded *perpetual* change as the fundamental law governing all things—a view which is summarized in his famous aphorism, 'You cannot step twice into the same river'. He also believed that there is a perpetual strife of opposites: hot and cold, wet and dry, and so on, are each necessary complements to the other and their eternal conflict is the very basis of existence. This world of change and conflict, however, is not just a chaos but is governed throughout time by a principle of order or balance of opposites, keeping them within their due bounds.

This principle was based on an idea that was accepted by other Greek thinkers of this period—the concept of Time as a judge. For example, Anaximander, in the only surviving fragment directly attributed to him, said that all things that are created must also perish, making atonement to one another for their injustice according to Time's decree. This idea was no doubt suggested by the cycle of the seasons with its alternating conflict of the hot and the cold, the wet and the dry. Each of these advances in 'unjust' aggression at the expense of its opposite and then pays the penalty, retreating before the counter-attack of the latter, the object of the whole cycle being to maintain the balance of justice. The fundamental assumption was that Time will always discover and avenge any act of injustice.

The concept of Time as a judge can also be attributed to the great Athenian statesman Solon (sixth century BC) who, according to Werner Jaeger, 'defends himself "before the bench of Time" '. (In this context it may be mentioned that in Athenian law courts it became the custom to have a clepsydra to ensure that most speeches were limited to half an hour.) This was an age when the state was being founded on the concept of justice. The original Greek word for 'justice', *themis*, signified 'divine law'. Although in the *Iliad* the word *dikē* denotes a judgement given by a judge or an assertion by a party to a dispute of his rights, in the *Odyssey* it signifies 'right' or 'custom'.[4] Later it became the slogan of those struggling for equal justice for all. Anaximander and Heraclitus extended the concept of justice to the whole universe:

In the life of politics the Greek language refers to the reign of justice by the term *kosmos*; but the life of nature is a *kosmos* too, and indeed this cosmic view of the universe begins with Anaximander's dictum. To him everything that happens in the natural world is rational through and through and subject to a rigid norm.[5]

Emphasis on the role of time characterized the Pythagorean idea of the *kosmos*. According to Plutarch, when asked what Time (*Chronos*) was, Pythagoras (sixth century BC) replied that it was the 'soul', or procreative element, of the universe. The extent to which Pythagoras and his followers may have been influenced by oriental ideas has long been a subject for argument. The Orphic idea of *Chronos*, which may have had an influence on Pythagoras, seems rather like the Iranian idea of *Zurvan akarana*. In particular, both were depicted as multi-headed winged serpents. Similarly, the dualism which played an important role in Pythagorean philosophy appears to echo the Zoroastrian cosmic opposition of Ohrmazd and Ahriman, although these two ultimates were regarded as personal gods and not as abstract principles like the Pythagorean ten basic pairs of opposites, such as limit versus unlimited, good versus bad, male versus female, odd versus even. The most fruitful feature of Pythagorean teaching was the key idea that the essence of things is to be found in the concept of number, which was regarded as having spatial and also temporal significance. Numbers were represented figuratively by patterns similar to those still found on dominoes and dice. Although this led to Greek mathematics being dominated by geometry, time was no less an important element in early Pythagorean thought. Indeed, even spatial configurations were regarded as temporal by nature, as is indicated by the role of the *gnomon*. This was originally a time-measuring instrument—a simple upright sundial. Later the same term was used to denote the geometrical figure that is formed when a smaller square is cut out of a larger square with two of its adjacent sides lying along two adjacent sides of the latter. Eventually, the term came to denote any number which, when added to a figurate number, generates the next higher number of the same shape (triangular numbers, square numbers, pentagonal numbers, and so on). The generation of numbers was regarded by the early Pythagoreans as an actual physical operation occurring in space and time, and the basic cosmogonical process was identified with the generation of numbers from the initial unit, the Monad, which may have been a sophisticated version of the earlier Orphic idea of the primeval World-egg.

It is well known that Pythagoras' belief in the significance of numbers was supported by his alleged discovery, with the aid of a stringed instrument, that the concordant intervals of the musical scale correspond to simple numerical ratios. This led many later Greek thinkers to regard musical theory as a branch of mathematics (together with geometry, arithmetic, and astronomy it constituted what eventually came to be called the *quadrivium*), although this view was not universally accepted, the most influential of those who rejected it being Aristoxenus of Tarentum (fourth century BC). He emphasized, instead, the role of sensory experience. For him the criterion of musical phenomena was not mathematics but the ear.

Long before the time of Aristoxenus, some of the most acute Greek thinkers had found that the concept of time was difficult to reconcile with their idea of rationality. Indeed, Parmenides, the founding father of logical disputation, argued that time cannot pertain to anything that is truly real. The essence of his difficulty was that time and change imply that the same thing can have contradictory properties—it can be, say, hot and cold, depending on the time—and this conflicted with the rule that nothing can possess incompatible attributes. His basic proposition was 'That which is *is*, and it is impossible for it not to be.' From this he argued that, since only the present 'is', it follows that past and future are alike meaningless, the only time is a continual present time and what exists is both uncreated and imperishable. Parmenides drew a fundamental distinction between the world of appearance, characterized by time and change, and the world of reality which is unchanging and timeless. The former is revealed to us by our senses, but these are deceptive. The latter is revealed to us by reason and is the only true mode of existence.

The difficulties involved in producing a logically satisfactory theory of time were emphasized by Parmenides' follower Zeno of Elea in his subtle paradoxes concerning motion. The most famous is the one generally known as the paradox of 'Achilles and the tortoise'. (The identification of Achilles' competitor as a tortoise is due to later commentators.) The tortoise is given an initial lead over Achilles, and the argument asserts that however fast Achilles runs he will never reach the tortoise. For, when Achilles reaches the point from which the tortoise starts, the tortoise will have advanced to a farther point. When Achilles reaches that point, the tortoise will be at a still farther point, and so on *ad infinitum*. Consequently, as Aristotle puts it in his account, 'the slower will always have a lead', in contradiction with experience, that is, the world of appearance. This argument assumes that space and time are infinitely divisible, but not all of Zeno's arguments involve this assumption. The problems that his paradoxes raise concerning the mathematical structure of space and time are still being discussed today.[6]

The difficulties discussed by Parmenides and Zeno do not occur if the concept of time is rejected as 'unreal'. Their influence on Plato (427–347 BC) is evident in the different treatment of space and time in his cosmological dialogue the *Timaeus*. Space exists in its own right as a given frame for the visible order of things, whereas time is simply a feature of that order. In Plato's cosmology the universe was fashioned by a divine artificer imposing form and order on primeval matter, which was originally in a state of chaos. This divine artificer was, in effect, the principle of reason, which by imposing order on chaos reduced it to the rule of law. The pattern of law was provided by an ideal realm of geometrical shapes which were eternal and in a perfect state of absolute rest, like the real world of Parmenides. Unlike the eternal ideal model on which it is based, the universe is subject to change. Time is that aspect of change which bridges the gap between the universe and its model, being 'a moving image of eternity'. This moving image manifests itself in the motions of the heavenly bodies. Plato's intimate association of time and the universe led him to regard time as being actually produced by the revolutions of the celestial sphere. A permanent legacy of his theory of time is the idea that time and the universe are inseparable. In other words, time does not exist in its own right but is a characteristic of the universe.

Plato's conclusion that time is actually produced by the universe was not accepted by Aristotle (384–322 BC), who rejected the idea that time can be identified with any form of motion or change. For, he argued, motion can be uniform or non-uniform and these terms are themselves defined by time, whereas time cannot be defined by itself. Nevertheless, although time is not identical with motion or change, it seemed to be dependent on them. He remarks that when the state of our minds does not appear to change we do not notice that time has elapsed. It is by being aware of 'before' and 'after' in change that we are aware of time. He came to the conclusion that time can be regarded as a numbering process associated with our perception of 'before' and 'after' in motion and change. He realized that the relation between time and change is a reciprocal one: without change time could not be recognized, whereas without time change could not occur. 'Not only do we measure the movement by the time, but also the time by the movement, because they define each other. The time marks the movement, since it is its number, and the movement the time' (*Physica*, iv. 220b). Aristotle recognized that movement can cease whereas time cannot, but there is one motion that continues unceasingly, namely that of the heavens. Clearly, although he did not agree with Plato, he too was profoundly influenced by the cosmological view of time. He rejected the identification of time with the circular motion of the heavens, but he regarded the latter as the perfect example of uniform motion. Consequently, it provides the perfect measure of time.

Although for Aristotle physics meant the study of motion and change in nature, the main emphasis was placed by him on the states between which change takes place rather than on the actual course of the motion itself. Thus the static form rather than the dynamic process became the characteristic concept in his philosophy of nature, and form and place were more fundamental than time. His natural philosophy was dominated by the idea of the permanence of the cosmos. He rejected all evolutionary theories and stressed instead the essentially cyclical nature of change.

Belief in the cyclical nature of the universe found its apotheosis in the concept of the Great Year, which the Greeks may have inherited from the Babylonians. The idea had two distinct interpretations. On the one hand, it was simply the period required for the sun, moon, and planets to attain the same positions in relation to each other as they had at a given time. This appears to be the sense in which Plato used the idea in the *Timaeus*. On the other hand, for Heraclitus it signified the period of the world from its formation to its destruction and rebirth. According to him the universe sprang from fire and will end in fire. This idea was probably transmitted from Iran, where it originated. The two inter- pretations were combined in late antiquity by the Stoics, who believed that, when the heavenly bodies return at fixed intervals of time to the same relative positions as they had at the beginning of the world, everything would be restored just as it was before and the entire cycle would be renewed in every detail. As Nemesius, Bishop of Emesa in the fourth century AD, later put it:

Socrates and Plato and each individual man will live again, with the same friends and fellow citizens. They will go through the same experiences and the same activities. Every city and village and field will be restored, just as it was. And this restoration of the universe takes place not once, but over and over again—indeed to all eternity without end. Those of the gods who are not subject to destruction, having observed the course of one period, know from this everything which is going to happen in all subsequent periods. For there will never be any new thing other than that which has been before, but everything is repeated down to the minutest detail.[7]

Nevertheless, as Ludwig Edelstein has pointed out, even in late antiquity there were philosophers, as well as historians and scientists, who regarded time as non-cyclical.[8] Cosmological recurrence involving the complete destruction of the universe and its exact re-creation, as believed in by the Stoic philosophers, must be distinguished from historical recurrence involving only the repetition of the general pattern of events, as believed in by the historian Polybius, for example.

Greek civilization not only gave rise to philosophy but it also produced, in the fifth century BC, the first real historians. Until then the Greeks believed that recent events were unimportant compared with the exploits of the heroes in Trojan times. Historiography arose when an event occurred which in its magnitude matched the greatest events celebrated in legend. The whole complex of events in the Persian wars from the fall of Sardis to the retreat of Xerxes was seen as a unity and formed what Robert Drews has called 'one Great Event of awesome proportions'.[9] Originally, the Greek historian's task was not to explain the present in terms of the past but to ensure that significant actions and events would not be forgotten in the future. Consequently, in its origins Greek historiography was more closely affiliated to epic poetry than to philosophy, and in its development it retained a commemorative function. Greek historians, for example Thucydides, tended to concentrate on the recent past, their object being to put in writing those significant actions which were remembered but had not yet been recorded.

The difficulties against which the 'fathers of history', Herodotus and Thucydides, had to contend were formidable. The Greeks of their time knew astonishingly little about their own past. Not only had they no documents going back more than a century or two, but much of what they 'knew' was merely myth and legend. Since their interest in the past was primarily moralistic, precise knowledge of actual events and when they happened was not required. Herodotus was able to establish some kind of time-sequence for the two centuries before his time, but he was a more diffuse writer than Thucydides, who was concerned with many events occurring in a shorter time interval. As Sir Moses Finley has pointed out, Thucydides, in writing about the Peloponnesian war (*History of the Peloponnesian War*, ii. 1), actually had to invent an appropriate system of dating, since each Greek city had its own calendar. In his time the year was usually indicated by the name of an official, for example at Athens the first archon and at Sparta the first ephor. Thucydides fixed the beginning of the war and dated subsequent events by counting how many years had elapsed from the start. Each year of the war he divided into two, which he called summer and winter, respectively. 'Simple enough,' Finley comments, 'yet the scheme was unique and the difficulties in making it work are nearly unimaginable today.'[10]

Whereas Herodotus transformed 'history' (*historia*) from a general enquiry about the world into an enquiry about past events, Thucydides believed that serious history could be concerned only with the present, or the immediate past. Although he did not succeed in imposing his strict standards of reliability on later Greek historians, he effectively discouraged the idea that one could do genuine historical research about the past.[11] Nevertheless, by the latter part of the fifth century BC there was a greater general awareness of the significance of time than there had been previously. Although Homer dealt with allegedly historical subjects, his was 'aristocratic' history, involving no chronology, no temporal continuity with later ages, and no real sense of the passage of time. For example, despite Odysseus' twenty years' absence from home, on his return neither he nor Penelope appear to have grown any older. In short, for Homer it made no difference that year follows year. On the other hand, by the time of Herodotus and Thucydides life in the *polis* did not consist of isolated episodes covering heroes but depended on the continuity of institutions, laws, contracts, and expectations. The passage of time had become more relevant.

In particular, problems of the calendar were the driving force that led to the initial development of Greek mathematical astronomy in the last decades of the fifth century BC. Most Greek religious festivals occurred at or near full moon, but since they were associated with agricultural activities they had to take place at the appropriate times of the year. It therefore became necessary to adopt a luni-solar calendar in which the months were measured by the phases of the moon but which also kept in step with the sun. Since the length of the lunar month is about twenty-nine and a half days and a calendar month cannot contain a fractional part of a day, it was arranged that the calendar months were alternately of twenty-nine and thirty days. As in Babylonia, this calendar was adapted to the sun by intercalating a thirteenth month from time to time, but this was left to local officials in the different cities to decide, and they did this individually and arbitrarily. Astronomers, on the other hand, sought to introduce a regular intercalation by means of a cycle of fixed period. According to Geminus, the author of a manual of astronomy of about 70 BC, the first such cycle obtained by the Greeks was an eight-year solar cycle containing ninety-nine months (three of which were intercalary), but there is considerable doubt about the origin of this cycle, known as the 'octaeteris'. The first well-attested cycle of this type was introduced in 432 BC by Meton. As previously mentioned (ch. 3), it was a nineteen-year solar cycle of 235 months (see Appendix 2). Astronomically based cycles such as the Metonic were used, however, only in scientific texts and had no influence on the various local civil calendars. Meton lived in Athens and appears as a character ridiculed by Aristophanes in *The Birds*, produced in 414 BC.

Besides historiography and mathematical astronomy another great innovation by the Greeks of the fifth century BC was the art of tragedy. Jacqueline de Romilly, Professor of Greek Literature at the Sorbonne, in her Messenger Lectures at Cornell University in 1967 on 'Time in Greek Tragedy', has argued that it was no coincidence that Greek tragedy was born at the same time as historiography. Tragedy involves the past, and it arose when the Greek awareness of time was becoming clearer and stronger. Greek tragedy concerns a single problem that becomes more and more urgent until it culminates in crime. A short continuous crisis, the origins and consequences of which cover a long period, seems to be the double requirement of tragedy and its double relationship to time. 'Its strength rests on a contrast between before and after; and the deeper the contrast the more tragic the event.'[12] Nevertheless, as Professor de Romilly makes clear, the Greeks disliked showing the action of time on moods and feelings. For example, when Euripides allows Iphigeneia to change her decision within a short time, Aristotle was shocked!

Contact with other nations (Egypt in the case of Herodotus) led to a greater awareness of the past, because of the evidence for long periods of time presented, for example, by the pyramids. Consequently many Greek writers of the fifth century and later realized that their own society was the end-product of a long period of advance. The more sophisticated Greeks were thus made to regard man in pre-Trojan times as much the same as his distant offspring, and this tended to demythologize the Greek legends, thereby placing the past in quite a new perspective.

After the fifth century, however, few, except writers on scientific subjects, had any belief in the idea of progress in the future. Indeed, the typical Greek tended to be backward-looking, since the future appeared to him to be the domain of total uncertainty, his only guide to it being delusive expectation. As for the philosophers, Plato thought that all progress consisted in trying to approximate to a pre-existing model in the timeless world of transcendental forms and Aristotle believed that it was the realization of a form which was already present potentially. Thus, for both of them the theory of forms excluded all possibility of evolution. Even in the sciences it was thought in later antiquity that all wisdom lay in the past. As E. R. Dodds has remarked, 'where men can build their systems only out of used pieces the notion of progress can have no meaning—the future is devalued in advance'.[13] Consequently, it is not surprising that in this period the main philosophical schools tended to reject the idea of progress and to hold cyclical views concerning the nature of time. Aristotle himself believed that the arts and the sciences have been discovered many times and then lost again. For example in the *Meteorologica* (339b27) he asserts: 'We must say that the same opinions have arisen among men in cycles, not once, twice, nor a few times, but infinitely often.'

Nevertheless, Arnoldo Momigliano has warned us that many Greek historians, as distinct from philosophers, paid little attention to the cyclical concept of time.[14] He has also pointed out that the future did not loom so large for them as it did for the Roman historians, who were anxious about the fate of their empire. Instead, the Greeks were more concerned with the present and the past. Writing *c*.40 BC, the historian Diodorus Siculus said of his predecessors:

Two views about the origin of mankind have been current among the most notable scientists and historians. One school, premising that the cosmos is ungenerated and indestructible, declares that the human race has always existed, and there was no time when it began to reproduce itself. The other holds that the cosmos has been generated and may be destroyed, and that men similarly first came into existence at a definite time.[15]

As regards the cyclical view, Momigliano says that the principal upholder of it in Greek historiography was Polybius (*c.*202–120 BC), but this opinion is based only on the constitutional chapters in his history of the world, for elsewhere he shows no sign of it. For example, he did not treat the Punic wars as repetitions of events that had already occurred in the past and would occur again in the future. His main theme was the increasing power of Rome in the Mediterranean and this, as Momigliano points out, provided him with a new historical perspective: 'Just because Fortune made almost all the affairs of the world incline in one direction, it is the historian's task to put before his readers a compendious view of the ways in which Fortune accomplished her purposes.'[16]

The concept of Fortune (i.e. Fate or Destiny) played a crucial role in later Hellenic thought, but different views were held. Aristotle criticized Democritus (*c.*460–390 BC) for believing only in efficient and not in final causes, that is, in strict determinism rather than in teleology. Aristotle believed that strict determinism must be rejected because it destroys the natural basis for distinguishing between voluntary and involuntary actions. For the purposes of law some actions must be regarded as voluntary, since only these can be justifiably punished. For Aristotle this argument was decisive. Similarly, although Epicurus (342–270 BC) unlike Aristotle accepted Democritus' atomism, he too rejected Democritus' belief in the strict determinism of all human actions. Instead of teleology, however, Epicurus advocated the existence of chance and free will, partly because, like Aristotle, he argued that you cannot blame or punish a man for something he cannot help doing, but also because he believed that there is a kind of spontaneity in men (and possibly in animals) that is manifested in our apparent freedom, to originate actions. Epicurus introduced the famous 'swerve' into the chain of strict causality, so as to account both for human free will and also for the existence of random motion in the universe; for otherwise all bodies would, in his opinion, fall with the same speed downwards. It was by stressing the chance element in destiny that he was led to the hedonistic philosophy of 'eat, drink and be merry for tomorrow we die!'

A very different point of view was advocated by the Stoics, beginning with Zeno of Citium (335–263 BC). Zeno and his followers rejected Plato's two-worlds theory of ideal forms and sense data. Instead, they believed in the organic unity of the whole universe, and they regarded intelligence as a refined material substance with a fiery nature. Unlike the Epicureans, the Stoics were strict determinists who advocated a philosophy of resignation in the face of worldly difficulties. For them Fate had a cyclical, or eternally recurrent, character. It was identified with Necessity and was symbolized by the unceasing rotation of a wheel, like the mythical wheel of Ixion. Since Fate was the power that kept order in the universe, as revealed particularly by the stars and planets, the prevalence of Stoicism influenced the growing belief in astrology in Hellenistic times and in the days of the Roman empire. The cyclical nature of events was regarded by many thinkers as inevitable, because it was thought that otherwise they would be deprived of both 'rationality' and 'legality'.

In late antiquity, both Plutarch (*c.* AD 46–120), the famous biographer and moral philosopher, and Alexander of Aphrodisias (*fl. c.* AD 200), an important commentator on Aristotle, criticized the views of the Stoics, as well as those of the Epicureans. Although he did not completely discard the astrological concept of destiny, Plutarch argued that there was a place in it for contingency. He formulated explicit definitions of 'necessity' and 'contingency' that are somewhat like the modern definitions of 'analytic' and 'synthetic': '*The necessary* is a possibility, the contradiction of which is impossible, but *the contingent* is a possibility, the contradiction of which is also possible.'[17] This distinction applies particularly to the future. Like Plutarch, Alexander of Aphrodisias argued that not everything is the product of inevitable destiny, since things that are produced by reason and by artists in the exercise of their craft 'do not seem to be produced by them through necessity, for they make each one of them indeed, but they are equally at liberty not to do so'.[18]

Just as there was no unique Greek idea of time, the history of the human race also presented itself to the Greeks in various forms. Besides the cyclical view and the progressive, there was the important tradition concerning a Golden Age in the remote past. The earliest extant account of this is to be found in the *Works and Days* of Hesiod (*c.*700 BC), who sought to account thereby for man's present condition and, in particular, for his need to work. According to Hesiod, in the 'good old days' before the lordship of Zeus when his father Kronos was king, there was a

Golden Age. Strictly speaking, Hesiod refers to a golden *race* rather than to the golden *age* of later writers. The idea of a primeval golden age can be traced back to the Sumerians (*c.*2000 BC). For them its most significant feature was freedom from feær. According to a Sumerian poet: 'Once upon a time there was no snake, there was no scorpion, / There was no hyena, there was no lion, / There was no wild dog, no wolf, / There was no fear, no terror, / Man had no rival.'[19]

According to Hesiod the age of idle luxury was followed successively by an age of heroes, a silver age, an age of bronze, and finally by the present iron age. Contrary to our knowledge today, this last was considered to be less civilized than the bronze age that preceded it. The original decline from the primeval Golden Age was explained by the myth of Prometheus, which has points of resemblance to the Hebrew myth of 'The Fall' described in Genesis. These include not only the creation of woman (Pandora corresponding to Eve) and the alleged evils that followed therefrom, but also the acquisition of 'forbidden knowledge', which in the Greek case included the discovery of fire.

By the classical period of Greek thought the myth of the Golden Age had partially given way to the opposite idea that man's early condition was 'nasty, brutish and short'. According to Moschion, who lived about the third century BC but wrote in the spirit of a century or two earlier, it was due to Time—'the begetter and nurturer of all things'—that 'The earth, once barren, began to be ploughed by yoked oxen, towered cities arose, men built sheltering homes and turned their lives from savage ways to civilized.'[20] Some writers on cyclical theories explicitly held out the hope that, although the world was in decline, the wheel would turn again so that eventually another Golden Age would repeat the idyllic conditions of the remote past. In his *Politics* Plato put forward a myth of cyclical change in which the creator imparts rotation to the universe and keeps it under his rule until, at the end of an era, he releases control. Thereupon the world reverses its rotation and everything starts to deteriorate until God reasserts his control and lets the universe rotate once again in the same direction as before. W. K. C. Guthrie, who has drawn attention to this myth, also points out that what Aristotle particularly deplored in the recurrent world catastrophes was the loss of accumulated knowledge and wisdom that they entailed.[21] Aristotle also doubted whether there could be time without thinking beings, since he regarded time as not merely succession but 'succession in so far as it is numbered', and nothing can be numbered unless there is someone to do the counting. The germ of this idea can be traced back to the sophist

Antiphon (*c.*480–411 BC), one of whose fragments contains the earliest Greek definition of time.[22] According to this definition, time has no substantive existence but is a mental concept or means of measurement—a point of view that strikes us today as being remarkably modern.

Finally, in surveying the role of time in ancient Greece, brief mention must be made of the instruments available for its measurement. Besides the gnomon, or sundial, and the clepsydra, or water-clock, an improved version of which with a more constant flow was invented by Ctesibius of Alexandria *c.*270 BC, there is evidence of more elaborate instrumentation, such as the 'Tower of the Winds' which can still be seen in Athens, north of the Acropolis. Designed and built by the astronomer Andronicus Kyrrhestes of Macedonia in the second quarter of the first century BC, with a wind vane and complicated sundials on each of its eight walls, its most interesting feature is a reservoir in a smaller building that stood next to its south side. Water from a nearby spring kept it filled. This is a requirement for water-clocks of the inflow type. With a constant head of pressure, the flow of water from a tap near the base of the reservoir could also be kept constant. The tap could be regulated so that the water flowing from it filled another tank in exactly twenty-four hours, while raising a float inside the tank a fixed distance. With a water-clock inside the tower and sundials outside, visitors could observe the time both by day and by night and also when the sky was cloudy as well as when it was clear. To cope with the traditional use of variable hours, so that the period of daylight always comprised twelve, Andronicus is thought to have used a system that is described in detail by his Roman contemporary, the architect Vitruvius. The clock's float was connected by a line to a counter-weight and this line passed round a horizontal shaft. As the float rose the shaft rotated and so did a circular metal plate attached to the end of it. On this plate was depicted a map of the heavens, and holes along the line of the ecliptic made it possible for a representation of the sun to be moved at intervals of a day or two in imitation of its annual motion. A complete rotation of the map every twenty-four hours simulated the daily rotation of the heavens. A grid of reference wires in front of the rotating map presented the hours, and as the solar image passed each wire it indicated the time as well as any sundial could. J. V. Noble and D. J. de Solla Price, who have described this Tower of the Winds in detail, believe that the interior must have been a dazzling sight. (Price has called the Tower of Winds 'a sort of Zeiss planetarium, of the classical world'.) They conjecture that Poseidon was a central figure between two fountains and that Hercules and Atlas held the wire grid before the bright disk which simulated the motion of the heavens. 'We live in an era in which we accept science and technology as commonplace,' Noble and Price conclude, 'and we expect them and our architecture to be efficient and functional. Athens . . . was a place of wonder and beauty, and it was a time to marvel at the achievements of mathematicians and astronomers—a time to build and admire a Tower of the Winds.'[23]

Ancient Israel

It has for long been held that our modern idea of time derives from that of early Christianity, which in turn can be traced back to that of ancient Israel and Judaism. Instead of adopting the cyclical idea of time, the Jews are said to have believed in a linear concept, based in their case on a teleological idea of history as the gradual revelation of God's purpose. Although there is much to support this view of the origin of our modern idea of time, it is now realized that it can only be adhered to with some reservations, as we shall see.

Following the Exodus from Egypt and the Settlement in Canaan in the latter part of the second millenium BC, the Jews found themselves in a region which was on the main line of communication between Egypt and Babylonia. Some time after the reigns of Saul, David, and Solomon the Jewish realm split into two. In 722 BC the northern kingdom, Israel, was overthrown and its capital destroyed by Sargon II. Two years later its people were deported to Assyria. In 586 BC the Babylonians destroyed Jerusalem, including the Temple, and many of the inhabitants of the southern kingdom, Judaea, were deported to Babylonia. According to Theodore Vriezen, Professor of Old Testament Studies at the University of Utrecht, the Babylonians deported mainly the upper classes and left perhaps 20,000 of the lower classes behind, so as not to let the country fall into total decay.[24]

The reaction of the Jews to these vicissitudes of fortune was profound. Appeal was made to the past for evidence of divine providence, current misfortune being explained as punishment for unfaithfulness to Yahweh, or God. It was believed that, if the nation were to become more zealous in its service of God, there would be more hope of deliverance. Although this was predicted to occur at some unspecified date in the future, belief in it was strengthened by the promise of a Messiah who would defeat Israel's enemies and restore the nation to its former glory. Consequently, the essential aim of the Jewish God in history was the salvation of Israel.

The definitive account of this belief was presented in the Book of Daniel, written long after the return from the Babylonian Exile under the stress of danger from the Seleucids just before the Maccabean rising in the second century BC. The appeal to the past was thus developed into a forward-looking philosophy of history. It has therefore frequently been maintained that for the ancient Hebrews time was a unidirectional linear process extending from the divine act of creation to the ultimate accomplishment of God's purpose and the final triumph, here on earth, of the chosen people, Israel.

According to the theologian O. Cullmann in his book *Christ and Time*, 'the symbol of time for primitive Christianity as well as for Biblical Judaism . . . is the *upward sloping line*, while in Hellenism it is the circle'.[25] On the other hand, the historian of political philosophy J. G. Gunnell has argued that, although the Hebrews were more oriented towards the future than the Greeks, who tended to look more towards the past, 'the concept of linear progression is a rationalization of the Hebrew experience of temporality'.[26] More recently another historian, G. W. Trompf in his study *The Idea of Historical Recurrence in Western Thought*, has drawn attention to the prevalence of what he calls 'notions of re-enactment' in the Old Testament which formed an ideological basis for the great Israelite festivals. Trompf also cites examples of re-enactment in Hebrew historiography, such as the crossing of the Jordan in Joshua, which was consciously likened to the traversing of the Red Sea in Exodus, and the similarity of the Babylonian exile to the earlier Egyptian bondage.[27] Moreover, even in eschatology and its presentation in terms of history, the idea of the future that dominated Hebrew thought involved a return to the primeval state that the Jews believed they had lost.[28] In other words, although they transferred their own Golden Age from the past to the future, a quasi-cyclical factor was involved.

Gunnell has pointed out that, unlike the Greeks, the Hebrews never tried to analyse the 'problem' of time as such. They seem neither to have conceptualized their experience of time nor formed an abstract idea of history. 'History was the space in which the drama of individual and social life unfolded according to the purpose of Yahweh, and cosmic time simply attested to the works of Yahweh and His power over the universe.'[29] One of the most significant features characterizing the Hebrew experience of time was the 'contemporaneity of past and future'. In other words, for the Hebrews the present was never a clearly delimited unit with precise boundaries but was part of a continuum stretching from the beginning to the end of time and was continually influenced by both past and future. It is significant that the Old Testament contains no numbered dates, despite its concern with an intricate historical record. The covenant was not just an important past event preserved by tradition but was the theme of what Gunnell calls 'a communal drama played out between Yahweh—He who will be there—and His people in time', for in the words of Deuteronomy 5: 3, 'The Lord made not this covenant with our fathers, but with us, even us, who are all of us here alive this day.'

The outstanding feature that distinguishes Hebrew thought from Greek thought (particularly that of Aristotle) was the idea of the cosmos as a creation of God that actually had occurred in history. In Hebrew thought, unlike Greek, nature was not divine, and God transcended all phenomena. The sun, moon, and stars were all God's creatures and served to show his handiwork (Psalm 19). Unlike the Egyptians and Babylonians, the Hebrews did not regard kingship as 'anchored in the cosmos'. In Hebrew religion, and in that religion alone, man was joined to God by a quasi-legal covenant, as a result of which the ancient bond between man and nature was destroyed.[30] Because of this, the Jews have sometimes been regarded as the 'builders of time', whereas the Greeks were the 'builders of space', the Romans the 'builders of empire', and the Christians the 'builders of heaven'. Eric Voegelin has emphasized the fundamental difference between what he calls the 'cosmological' civilizations, which presupposed the political symbolization of the cosmos typified by Babylonia with its epic of Marduk, and 'eschatological' civilizations such as the Hebrew—but first exemplified by the Iranian—based on the religion of Zarathustra.[31] Although the main emphasis in Jewish eschatology has always been on the fate of the nation, the doctrine of personal immortality (which originated with Zarathustra's passionate belief in the justice of God) seems to have been adopted by the Jews during or after their Babylonian exile. Belief in this doctrine was the greatest innovation of post-exilic Judaism. Irrespective of the precise role of linearity in the Hebrew notion of time, it was for long assumed that the eschatological nature of that concept greatly influenced, by way of Christianity, the development of our modern idea of time's unidirectional non-cyclic nature.

In recent years, however, there has been a growing tendency to question the assumption that, prior to the advent of Christianity, Israel was unique among the nations of antiquity in the significance it attached to history and the non-repeatability of events. For, not only did the explicit recognition of forward movement in time and the rejection of the idea of endless recurrence originate with Zoroastrianism, but in the last twenty years or so Old Testament scholars have drawn attention to the similarity between some passages in the Old Testament and certain Mesopotamian texts. As a result, J. Van Seters, Professor of Biblical Literature in the University of North Carolina, and others have argued that the idea of there being a unique 'divine plan of history' in the Old Testament has been 'greatly overstated'.[32] In other words, the conviction of the Israelites that they were 'God's chosen race' is now generally regarded by scholars as not greatly different from the fundamental belief on which the Sumerian and Babylonian city-states had been based, namely, that the king was divinely elected.

The Hebrews were influenced by the Sumerians and Babylonians in other ways too, including the measurement of time. Consequently, their calendar was based on the moon. As among other peoples who count by lunations, the Hebrew month begins when the moon's slim crescent is first visible in the evening twilight. As early as the time of Saul the festival of the new moon was celebrated with great solemnity. Later, when Jerusalem was the capital, as soon as the appearance of the new moon had been proved by credible witnesses before the Sanhedrin, messengers were dispatched from there to announce the commencement of the new month.

Originally the Jewish year commenced at the autumn equinox. The Jewish civil year still begins at this time, but since the exodus from Egypt the Jewish ecclesiastical year has begun with the month Nisan at the spring equinox. By using years of different lengths, depending on the insertion or not of an intercalary month, reasonable agreement with the sun was maintained. Not only the new moon but the full moon too was regarded by the Hebrews as being of great religious significance, and the timing of Passover was determined by the first full moon on or after the spring equinox. As regards the numbering of their years, the Jews used the same era as the Seleucids of Syria from the time they came under their rule, in the second century BC, until the destruction of the Temple by the Romans in 70 AD.

In some of the older parts of the Bible, particularly in those concerning the earlier prophets, the moon is frequently mentioned in connection with the Sabbath, which commemorated the seventh day of creation, when the Lord rested from his labours. In his book *Rest Days* Hutton Webster has drawn attention to the passage in 2 Kings 4: 23, describing how, when the Shunammite woman wanted to go to the prophet Elijah to beg him for her son's life to be restored, her husband objected, saying 'Wherefore wilt thou go to him today? It is neither new moon nor Sabbath.' As Webster goes on to point out, when study of the cuneiform records revealed that the Babylonian *shabbatum* (full-moon day) also fell on the fourteenth (or fifteenth) day of the month, we were presented with another survival of what must have been the primary meaning of the Hebrew term *shabbath*.

Although the Hebrew seven-day week ending with the Sabbath (the only day to which a name was given) resembles the Babylonian seven-day period ending with an 'evil day', there are imporant differences. For the latter cycle was always directly associated with the moon, whereas the Hebrew week was not but was continued from month to month and year to year regardless of the moon. Moreover, the Babylonian evil day was observed only by the king, priests, and physicians, whereas the Hebrew Sabbath was observed by the whole nation. As Webster points out, 'To dissever the week from the lunar month, to employ it as a recognized calendrical unit, and to fix upon one day of that week for the exercises of religion were momentous innovations which, until evidence to the contrary is found, must be attributed to the Hebrew people alone.'[33]

By the time that Israel became part of the Roman empire, the idea was already widespread among the various religious sects that the 'End of the World' was at hand, although it was only for the Essenes of Qumran that this belief assumed a definite form; that, whereas the First Judgement at the time of Noah had been destruction by water, the Last Judgement would be destruction by fire. The Essenes, an extremely ascetic sect who withdrew to the desert region of Judaea near the Red Sea, appear to have originated in the middle of the second century BC at the time of the Maccabean revolt against the misguided Hellenizing reforms of the Seleucid ruler of Palestine Antiochus IV. He had recently taken over this region, which had previously been subject to the Ptolemies in Egypt. The Essenes were not only greatly inclined to apocalyptic views and legalism, but they were frantically anti-Hellenistic. Following the conquests of Alexander, Egypt and the rest of the area that we now call the Middle East were dominated by Hellenistic customs and views. Greek became the *lingua franca* of this region and remained so during the time of the Roman empire. Because of this, the books of the New Testament appeared in Greek, although the language spoken by Jesus and his disciples had been Aramaic.

Many Jews, especially outside Palestine, became Hellenized, but inside that country only the Sadducees were sympathetic to Hellenic culture. As the most intellectually enlightened of the sects, they were on the side of the ruling power. Their name signified 'sons of Zadok', the high priest at the time of David who was thought to be a descendant of the younger son of Aaron. They were the main enemies of the predominant sect, the Pharisees, who believed that salvation would only come if they adhered strictly to the Mosaic law, as originally set out in Deuteronomy where it was made clear that the chosen people must be a 'clean' people. The Deuteronomic standpoint was later canonized in the Torah and in the books of the prophets and became central in Jewish life. Indeed, of the books of the Old Testament so far found to have been in use among the Qumran sect, with the exception of Isaiah and the Psalms, most are copies of Deuteronomy.[34]

It is of particular interest to us today that the rivalry of Pharisees and Sadducees extended to their differing views concerning the way time should be measured. For, whereas the Pharisees adhered to the lunar year (with intercalary months so that the agricultural year kept pace with the sun), the Sadducees adopted the luni-solar year used by the Greeks. Each sect accused the other of wishing to observe the prescribed religious festivals at the wrong times, although in practice they had to keep to the same dates. Because the Pharisees were the predominant sect, few Sadducee writings have survived. Among those that have, particular interest attaches to the Book of Jubilees, which was probably composed about the year 110 BC.[35] The basis of this calendar of jubilees seems to have been the famous Pythagorean right-angled triangle of sides three, four, and five.[36] The sum of the first two gives the number of days in the week, the sum of all three gives the number of months in the year, and the sum of their squares gives the number fifty. According to Philo of Alexandria, a first-century Graeco-Judaic philosopher who wrote many works that still survive, including commentaries on the Old Testament, fifty was regarded as the holiest of numbers and 'the principle of the generation of the universe' (*De vita contemplativa*, 65). The significance of a fifty-year cycle in Jewish life, with the remission of debts, the release of slaves, etc., was eventually responsible for the practice that has been followed by successive Popes, since 1300, in declaring a Jubilee of the Roman Church every fifty years.

Imperial Rome and early Christendom

Because of the way in which it began, Christianity inherited the peculiar Jewish view of time with its hope of redemption from successive oppressors. At first Christians looked upon the risen Jesus as the Messiah whose return was imminent and would bring to an end the existing world-order. Gradually, as time passed without this return occurring, Christians had to cope with a world that continued to exist, its end being postponed to an indefinite future. If Jesus were the Messiah, then he had already come and a new interpretation was necessary. The birth of Jesus thus came to be regarded as dividing time into two parts, because it ended the first phase of the divine purpose and initiated the second. Unlike adherents of other contemporary religions in the Roman empire, except Judaism, Christians regarded their religion as expressing the purpose of God in history; but whereas Judaism was concerned primarily with the fortunes of Israel, Christians considered their faith to be of universal significance. The crucifixion was considered by them to be a unique event not subject to repetition. Consequently, time must be linear rather than cyclic. This essentially historical view of time, with its particular emphasis on the non-repeatability of events, is the very essence of Christianity. It is brought out clearly, and even contrasted with the Hebrew view, in the Epistle to the Hebrews, 9: 25–6: 'Nor yet that he should offer himself often, as the high priest entereth into the holy place every year with the blood of others; For then must he often have suffered since the foundation of the world; but now once in the end of the world hath he appeared to put away sin by the sacrifice of himself.'

The world in which Christianity originated was that of the Roman empire. The age was one in which a variety of religions flourished, many of oriental origin. In general it was an extremely superstitious age. On many days of the year the traditional religious calendar forbade business of any sort. In particular, on days of ill-omen ships could not set sail. Thus, no Roman skipper would move off from a port on 24 August, 5 October, or 8 November, and it was thought bad to be at sea at the end of the month.[37]

As Sir Ronald Syme has pointed out, the Romans had a special veneration for authority, precedent, and tradition, and they greatly objected to change unless it was thought to be in accord with ancestral custom, which meant in practice the sentiments of the oldest living senators. The Romans tended to be suspicious of novelty, and the word 'novus' had for them a sinister ring, although their memory of the past reminded them that change had often come about, although at first resisted. As Syme has remarked, 'Rome's peculiar greatness was due not to one man's genius or to one age, but to many men and the long process of time.'[38]

One of the main inspirations of Roman historians was the cult of ancestors and the propensity of noble families to commemorate their deeds. Unlike the Greek historians, they made it their business as patriots to present a comprehensive survey of their country's past. The first history of Rome was, however, written by a Greek, Polybius, who lived in Rome in the second century BC. Later historians of note include Caesar, Sallust, Livy, Tacitus, and Suetonius. Livy (*c*.59 BC–AD 19) gave his history the title *Ab urbe condita* (*'From the Foundation of the City'*) and began with Aeneas. Omens and prodigies abound in his work, so that compared with him Herodotus seems almost modern. The only Roman historian who can be compared to Thucydides is Tacitus (*c*. AD 55–117). Both were great stylists and for them history was the ultimate tribunal before which the actions of rulers and others can be judged, 'but where Thucydides was a magistrate, Tacitus was an advocate—the most brilliant, perhaps, who ever sought to determine the judgement of Time, but an advocate all the same'.[39] Although much of his historical writing depended on oral testimony, he is of all ancient historians the one who most frequently cites the authors and documents that he has consulted. He had an exalted idea of history that is well illustrated by his claim (*Annals*, iii. 65) that the historian's duty is 'to rejudge the conduct of men, that generous actions may be snatched from oblivion, and that the author of pernicious counsels, and the perpetrator of evil deeds may see, beforehand, the infamy that awaits them at the tribunal of posterity'. The Romans tended to regard the course of history as alternating between defection from and adherence to traditional values. As E. R. Curtius has pointed out, the pious attitude of the Romans to their past and their tendency to regard it as if it were a part of the present signified a kind of timelessness that excluded a genuinely historical view of the world and was very different from our sense of temporal perspective.[40]

Although the Romans respected the literary and other cultural achievements of the Greeks, they were puzzled by the importance assigned by them to mathematics. The outstanding exception to the general conclusion that the Romans were not really interested in science was Lucretius (*c*.94–55 BC), whose *De rerum natura* is nowadays regarded as the greatest philosophical poem ever written. Although it impressed both Cicero (106–43 BC) and Virgil (70–19 BC), the Epicureanism on which it was based made little impression on the Romans, except for its hedonistic aspect. As regards the concept of time, the poem is remarkable for its modern point of view: 'Similarly, time by itself does not exist; but from things themselves there results a sense of what has already taken place, what is now going on and what is to ensue. It must not be claimed that anyone can sense time itself apart from the movement of things or their restful immobility.'[41]

Unlike Epicureanism, Stoicism had a considerable appeal for the more educated citizens. A famous passage in Virgil's *Fourth Eclogue* gives expression to the concept of the 'Eternal Return': 'Now is come the last age of the song of Cumae; the great line of the centuries begins anew . . . A second Tiphys shall then arise, and a second Argo to carry chosen heroes; a second warfare, too, there shall be, and again shall a great Achilles be sent to Troy.' The stoical attitude of philosophical resignation replaced the old Roman polytheism which had become more and more a meaningless formality. In so far as Jupiter survived he was the personification of Providence or Destiny. The deification of the Emperors, introduced by Augustus, was not taken too seriously and signified little more than in later ages was implied by the adjective 'Holy' in the title 'Holy Roman Emperor'.

Although for the upper classes the *Pax Romana* in the age of the Antonines (second century AD) came as a great opportunity to concentrate on and uphold the customs of their local town or district, for humbler men it provided wider horizons and unprecedented opportunities for travel. As a leading authority on late antiquity has pointed out, 'merchants were constantly on the move, seeking opportunities in the underdeveloped territories of Western Europe, often settling far from their native towns'.[42] Indeed, one merchant from Phrygia is known to have visited Rome no fewer than seventy-two times. This new freedom to travel safely far and wide had a profound effect not only on men's lives but also on their thoughts and beliefs; these men who were being uprooted provided 'the background to the anxious thoughts of the religious leaders of the late second century'.[43] It was from them and no longer from the humble and oppressed, as in the previous century, that the converts to Christianity were now mainly recruited.

At this time, however, Christianity was only one among a number of competing religions in the Roman empire which were coming increasingly under the cosmopolitan influence of Hellenistic civilization. Stoicism had declined, the last prominent exponent of its philosophy being Marcus Aurelius who ruled from 161 to 180. His *Meditations*, with their emphasis on the vicissitudes of perpetual change, exude an air of world-weariness. The following century saw the spread of Gnosticism, the believers in which laid claim to secret, or privileged, knowledge and so were called *gnostikoi* ('knowers'). It was a way of thought based on the general Hellenic idea that salvation is obtained by knowledge. Besides Christian forms there were others such as Hermeticism and Manichaeism. One of the most characteristic features of Gnostic thought was the fundamental dualism of God and the world, the Deity being regarded as completely transcending the world which, so far from being his creation, was the realm of the Devil and consequently irredeemably evil. Gnosticism can be looked upon as a revolt against Greek science. Although dualistic it was quite different from Platonism: cosmic time was not the moving image of eternity but 'at best a caricature of eternity, a defective imitation far removed from its model'.[44] Similarly, Gnosticism was opposed to orthodox Christianity by its hostility to history, for instead of being based on the idea that God prepares for the future by way of the past it regarded the world as one from which God was absent. Consequently, when Gnosticism was combined with Christian ideas the result was soon judged unacceptable by the Church. Nevertheless, this peculiar combination had a long life and was destined to reappear in the Middle Ages as the Albigensian heresy that flourished for a while in southern France but was eventually crushed in the first quarter of the thirteenth century by the northern French at the command of the most powerful of the medieval Popes, Innocent III.

Among other forms of religion that flourished in the Roman empire was Mithraism. This extremely masculine religion appealed to the Roman army. The founding father of modern Mithraic studies, Franz Cumont, showed that Roman Mithraism was a continuation of the Iranian religion of Zarathustra and that its origins can be traced back to the Hindus, for in the Vedic hymns we encounter the name Mitra. According to Cumont, despite the theological differences between the Vedas and the *Avesta*, 'the Vedic Mitra and the Iranian Mithra have preserved so many traits of resemblance that it is impossible to entertain any doubt concerning their common origin'.[45]

Two different iconographical images of Mithra have survived. In one found only in the West, e.g. in the course of excavations in the City of London, he appears as a handsome bull-slaying god, signifying the renewal of the world at the time of the New Year. In a marble group in the British Museum depicting the bull-slaying Mithra the most striking feature is that three spikes of wheat are shown issuing from the wound of the sacrificed bull.[46]

Mithra's other form, found in the Eastern as well as the Western world, is as a lion-headed monster around whose body a serpent is coiled. The snake is sometimes decorated with signs of the zodiac. It therefore represents the path of the sun around the ecliptic and indicates the connection between Mithra and the Iranian god of time, Zurvan. This symbolism is similar to that found in many ancient cultures, including those of Mesoamerica, in which the serpent represents cycles of endless time, perhaps suggested by the fact that the snake periodically sheds and renews its skin. In the story of the Fall in the third chapter of Genesis the destroyer of Man's primeval innocence is also depicted as a serpent. The representation of endless time by a snake swallowing its own tail and bearing the legend 'My end is my beginning' occurs later in rings worn on the finger, such as that possessed by Mary, Queen of Scots.

The Mithraic lion-headed god symbolized Eternity. This representation of Mithra appears to have been derived from Egyptian art, and it has been suggested that it may have been influenced by the bandages of mummified corpses.[47] There are indications that in Egypt the god of eternal time was identified with Osiris. In some representations, in the *Book of the Dead*, the phoenix is depicted as arising from him.[48] The phoenix is also sometimes depicted in Mithraic contexts. Since Egyptian theology was influential in Imperial Rome, M. J. Vermaseren has argued that 'neither Iran nor Egypt alone formed the cult of the lion-headed god in Mithraism, but the Hellenistic age in general, of which Egypt was a major component, formed a concrete representation of the abstract idea of eternity.'[49]

Besides the various religions of eastern origin that flourished in Rome during the second and third centuries AD, there was also a resurgence of philosophical speculation. This was based on a revival of Plato's ideas and so is called Neoplatonism. The greatest figure of this school was Plotinus (*c.*205–70). Born in Egypt, he settled in Rome in 244. In his philosophy reality is the spiritual world contemplated by reason, the material world being a mere receptacle for the ideal forms imposed on it by the world-soul. The seventh part of his third *Ennead* ('On Time and Eternity') can be regarded as meditation on the passage in Plato's *Timaeus* (37–8) where time and the creation of the world are discussed.[50] Plotinus believed that the origin of time was to be found in the life of the world-soul. The question as to whether time could conceivably exist if there were no 'soul' (or mind) to apprehend it had been raised, but not answered, by Aristotle, whose definition of time as the 'numbering' of motion and change in relation to before and after appeared to presuppose the existence of a 'soul' that contemplates and measures it. For most philosophers of classical antiquity the world was both animate and divine. Consequently, it was possible for them (but not for Christians, because they rejected pantheism) to speak of a world-soul that could measure time, and this was, in fact, the answer given by Plotinus to Aristotle's question. Plotinus also advanced beyond Plato by modifying the latter's famous metaphor of time as the moving image of eternity, since he was more concerned to stress the difference between, rather than the resemblance of, time and eternity. In his opinion, although everything that exists must be like its cause, the fact that one thing is produced by another implies that they are different. Adopting a hierarchical stand-point and preferring to speak in terms of 'life' rather than 'motion', Plotinus regarded time as an intermediate between eternity (or the higher soul that contemplates eternity) and the motion of the universe which reveals time as the 'life' (or creative power) of 'soul'.[51] Although not a Christian, Plotinus was in some respects a forerunner of St Augustine, particularly because he thought of time in psychological terms.

Early in the fourth century the struggles that had occurred inter-mittently between the Roman state and the Christian Church ended with the latter proving the stronger, partly as a result of the military upheavals that had threatened the former in the middle of the previous century. Two events of outstanding importance then helped to settle the fate of each; the capital of the Roman empire was transferred to Byzantium, renamed Constantinople, and Christianity became the state religion. In his earlier years the emperor Constantine (*c.*288–335) had been an adherent first of Hercules and then of Sol Invictus. His conversion to Christianity marked a turning-point in the history of both the Church and Europe. The 'Chi–Rho' monogram of Christ began to appear on Constantine's coins in the year 315. At the same time, the Bishop of Rome began to become more important in the west, partly because the Emperor no longer lived in the old capital.

Whereas Augustus had had the poet Virgil to sing his praises, Constantine had the ecclesiastical politician and historian Eusebius as the man who sat immediately to the right of his throne during the sessions of the Council of Nicaea in 325 and exercised a decisive influence on the creed and discipline of the Universal, or Catholic, Church. Constantine was declared to be Emperor by divine right. As a result, he 'gained rather than lost by his willingness to exchange the style and title of a god for that of God's vice-gerent'.[52] Later that century the Empire was finally split up into an eastern and a western part. Thereafter, the latter, which was the more directly threatened by invaders, could no longer call upon the stronger military forces of the former. After the western ruler

Honorius had refused the province of Noricum (southern Austria) to the Visigothic king, Alaric, whose lands were under pressure from the Huns to the east, the latter marched with his troops on Rome in the year 410 and sacked the 'Eternal City'. This unprecedented catastrophe shocked the Empire profoundly. It led the Bishop of Hippo (near Carthage) to write soon afterwards his great book *The City of God*, the first philosophy of history, in order to rebut the charge that the sack of Rome was punishment for the abandonment by its citizens of their traditional pagan gods.

Like Paul of Tarsus, Augustine of Hippo was a convert to Christianity, having previously been a Manichee and then a Neoplatonist like Plotinus. His *Confessions*, written not long before the Fall of Rome, was an even more original form of literature than Rousseau's written more than a thousand years later, for it was the first true autobiography. It led William James to call St Augustine, although he lived so long ago, 'the first modern man'. In it he gave an account of his life including his conversion to Christianity and his struggle against rival doctrines.

Even after he had ceased to be a Neoplatonist, St Augustine remained very much under the influence of Plato's philosophical ideas, in particular those concerning time. Like Plato, he believed that the concepts of time and the universe were inseparable, each being essential to the other. In *The City of God* (xi. 5, 6; xii. 16) he argued that time can have no existence unless things are actually happening, and in his *Confessions* (xi. 14) when replying to the question what was God doing before he made heaven and earth, 'I answer not,' he wrote, 'as one is said to have done merrily (eluding the pressure of the question), "He was preparing hell (saith he) for pryers into mysteries".' In both books we find him passionately concerned with the nature of time and vigorously rejecting cyclical theories of history. In *The City of God* (xii. 13) he wrote:

The pagan philosophers have introduced cycles of time in which the same things are in the order of nature being restored and repeated, and have asserted that these whirlings of past and future ages will go on unceasingly. . . . From this mockery they are unable to set free the immortal soul, even after it has attained wisdom, and believe it to be proceeding unceasingly to false blessedness and returning unceasingly to true misery. . . . It is only through the sound doctrine of a rectilinear course that we can escape from I know not what false cycles discovered by false and deceitful sages.

Like Plotinus before him, St Augustine, in Book XI of his *Confessions*, submitted Aristotle's concept of time to searching criticism. He argued that time and motion must be more carefully distinguished from one another than they were by Aristotle. In particular, he objected to correlating time with the motions of the heavenly bodies, since time would still exist if the heavens should cease to move but a potter's wheel continued to rotate. For there would be some temporal duration represented by each revolution of the wheel and a certain number of these revolutions would still take place in the interval of time we call a day, even though the motion of the sun had ceased. Similarly, if a body be sometimes in motion and sometimes at rest, we measure its period of rest as well as its period of motion by time. In place of Aristotle's association of time with motion and his appeal to the uniform daily revolution of the heavens as its basis, St Augustine turned, not as Plotinus had done to the concept of the 'world-soul', but to the human mind for the ultimate source and standard of time. Whereas Aristotle did not enquire into the mental process by which we perceive time, because he believed that our minds must necessarily conform to the time of the physical universe, St Augustine took the mind's activity as the basis of temporal measurement. He considered the problem of measuring the time taken by a voice in making a single sound. Clearly, before the sound begins we cannot measure the time it is going to take, but after it has sounded how can we measure it, since it is then no more? Nor can we measure it in the present if we regard the present as an indivisible instant that is truly momentary and without duration. St Augustine came to the conclusion that we can measure time only if the mind has the power of holding within itself the impression made by things as they pass by even after they are gone. In other words, we do not measure the things themselves but rather something that remains fixed in the memory. It is the impression that passing events leave in the mind that we measure, for only this impression remains after they have passed. The mind has the power of distending itself into the future by means of anticipation and the past by means of memory. In the present there is only the attention of the soul by means of which the future becomes the past, and only when the constant diminishing of the future of the sound has made it entirely past can the mind measure it in terms of some preconceived standard. St Augustine did not explain how the mind could be an accurate chronometer for the timing of external events, but as the pioneer of the study of psychological time he stands in the front rank of those who have contributed to the understanding of our sense of time.

Whereas for most Greeks and Romans, whether they believed in cycles or not, the dominant aspects of time were the present and the past, Christianity directed man's attention to the future. In the words of the philosopher Erich Frank, 'With Christianity . . . man acquired a new understanding of time.'[53] The Christian view of time directed to the future, as presented by St Augustine, differed from the ideas of time current in Classical antiquity in that it was neither cyclic nor would it continue indefinitely without anything essentially new occurring. John Baillie has made the further point that, in his detailed criticism of cyclical views of time, St Augustine was anxious to defend the doctrine of creation and particularly its corollary that 'through the creative power of God the course of events is characterized by the emergence of genuine *novelty*.'[54] In assessing the importance of St Augustine for the development of the Christian view of time, his writings can be contrasted with the New Testament. Olaf Pedersen has recently drawn attention to St Paul's complete indifference to time and chronology: he never even dated his letters.[55] Presumably this total lack of interest was due to his belief, which he shared with other early Christians, that the Second Coming was imminent (Romans 13: 11–12). Time for Christians began with the Creation and would end with Christ's Second Coming. World history was bounded by these two events. The spread of this belief marks the divide between the mental outlook of Classical antiquity and that of the Middle Ages. Moreover, our modern concept of history, however rationalized and secularized it may be, still rests on the concept of historical time which was inaugurated by Christianity.[56]

Although it is to Christianity that we owe our modern temporal orientation, it is to the Romans that we are mainly indebted for the form of our calendar and conventions of time recording. Prior to Julius Caesar, however, Roman achievements in chronometry were far from impressive. For example, when Rome's first sundial was brought to the city from Sicily in 263 BC, during the first Punic war, and was erected in the Forum it was inaccurate because it indicated the time appropriate to the place whence it came which was more than four degrees to the south. It was not until 164 BC, almost a century later, that a public sundial was erected that was appropriate to Rome's latitude. A public clepsydra was set up in Rome in 158 BC by Scipio Nasica. The introduction of clocks into Roman law courts, following the practice in Greece, led some unscrupulous lawyers to bribe the clepsydra attendant to regulate the water supply in their favour. From Caesar we learn that water-clocks were used in military camps to time the night watches (*De bello Gallico*,

v. 13). According to St Mark (13: 35), there were four night watches: evening, midnight, cock-crow, morning.

Writing in Imperial times, the poet Juvenal (*c.* AD 50–130), informs us that in his day wealthy members of the upper class had private water-clocks and special slaves to read them and announce the hours to their masters. Clocks thus came to be regarded as status symbols. An example of this occurs in Petronius's *Feast of Trimalchio*, Trimalchio having a beautiful clock in his dining room. Nevertheless, the unequal hours and comparative inaccuracy of Roman clocks led Seneca (*Apocolocyntosis*, ii. 2–3) to complain that it was impossible to tell the exact hour 'since it is easier for philosophers to agree than for clocks'!

Our present calendar is a modification of the calendar introduced by Julius Caesar on 1 January 45 BC and since named after him (see Appendix 1). Previously, the Romans had tried to bring their civil calendar, which like many ancient calendars was based on the moon, into line with the astronomical year based on the sun by adopting a system involving an additional or intercalary month every second year. Since the length of this month was not determined by any precise rule, the pontiffs were left to exercise their discretion, and they frequently abused this power for political ends. By manipulating the number of days in the intercalary month they could prolong a term of office or hasten an election, with the result that by the time of Julius Caesar the civil year was about three months out of phase with the astronomical year, so that the winter months fell in the autumn and the spring equinox came in the winter.

Acting on the advice of the Greek astronomer Sosigenes, Caesar directed that to correct this anomaly the year 46 BC should be extended to 445 days. Although this led to it being called 'the year of confusion', his object was to put an end to confusion. He also abolished the lunar year and the intercalary month and based his calendar entirely on the sun. He fixed the true year at 365¼ days and introduced the leap year of 366 days every fourth year, the ordinary civil year comprising 365 days. He ordered that January, March, May, July, September and November should each have 31 days, the other months having 30 days, except February which should normally have 29 but in leap years would have 30. Unfortunately, in 7 BC this neat arrangement was interfered with in order to honour Augustus by renaming the month Sextilis after him (he believed that it was his lucky month) and assigning to it the same number of days as the preceding month that had been renamed after his murdered great-uncle by Mark Antony. A day was thus taken away from February and transferred to August. To avoid having three months each of 31 days occurring in succession, September and November were each reduced to 30 days and October and December were each raised to 31. Thus to honour the first of the Roman emperors an orderly arrangement was reduced to an illogical jumble that many people find difficult to remember but which in the course of 2,000 years has been successfully imposed on most of the world.

Although originally the Roman calendar began in the spring on 1 March (as reflected in our names for the months September to December), the consuls, who were elected for one year, in 153 BC began to take office on 1 January. From then on the year was regarded by the Romans as beginning on that day. Later this choice was considered to be pagan by the Church because of the festivities traditionally associated with it. Instead the Church preferred to use the Annunciation for the first day of the year, and this led to the adoption of 25 March, nine months before Christmas, although this choice was by no means universal. (Astronomers, as a rule, kept to 1 January as the beginning of the year. Generally, the history of the beginning of the civil year is complicated.[57] For example, in Venice the year began on 1 March until the fall of the republic in 1797.) From AD 312 'indiction cycles' of fifteen years' duration were introduced by the Emperor Constantine for taxation purposes and led to the Byzantine year being reckoned from 1 September, the date on which each year of an indiction cycle began. They remained popular in the West throughout the Middle Ages and even continued to be used by the supreme tribunal of the Holy Roman Empire until its abolition by Napoleon in 1806.

The Romans made use of the idea of denominating the years by a single era count. This idea had been put into practice in 312/311 BC by Seleucus I, the Hellenistic ruler of Babylonia. The following century the Greek system of dating by successive Olympiads from the first in 776 BC was begun, either by the historian Timaeus of Sicily or by Eratosthenes, the famous librarian of the Museum in Alexandria and measurer of the earth, and later Greek chronology was based upon it. The Roman system of dating *ab urbe condita* (i.e. from the foundation of Rome) was introduced by Varro in the first century BC and was based on the date assigned to the fabled founding of the city. Although this system was ratified by Julius Caesar in 46 BC and was widely used, there was some uncertainty about the precise relation of the resulting Roman dates to those of the Olympiads.[58] According to the historian Polybius, the founding of Rome occurred at an Olympiad dating corresponding to 750 BC.

Other dates were also ascribed to this event. In the time of Augustus the list that was compiled of magistrates of the Republic was based on counting from 752 BC. The date that was eventually generally accepted was 753 BC, originally suggested by Varro (116–27 BC). According to tradition, the birthday of Rome was on the festival of Parilia, 21 April. Consequently, on that day in the year AD 247 the Romans celebrated the thousandth anniversary of the founding of their city. Coins were minted bearing the famous inscription *Roma aeterna*—'Rome, the eternal city'.

Among the conventions for the division of time that have come down to us from Imperial Rome is the seven-day week. Its origin can be traced back to the Sumerians and Babylonians. It was never used by the Greeks, who divided the month into three parts of ten days each, but it was employed by the Jews (see p. 55). Originally, the Romans had a complicated system of dividing the month, with Calends (from which our word 'calendar' is derived) on the first, Ides on the fifteenth of March, May, July, and October and on the thirteenth day of the other months, and Nones occurring eight days before the Ides. Originally, the Calends were the days of new moon and the Ides the days of full moon. Initially, the year was divided into the ten months March to December, the period midwinter to spring being left out because there was little agricultural work to be done then. Later this period was divided into the months January and February. In the early history of Rome the only times recognized in the daylight period were sunrise, midday, and sunset. The nights, however, were divided into four *vigilae* (or watches), this system being presumably of military origin. The days were counted backwards, from the Calends, Nones, and Ides, respectively. The day from which the Romans calculated and the day to be designated were both included, for example, 2 January was designated *ante diem IV Non. Jan.* The Nones were so named because they occurred on the 'ninth' day before the Ides. The days after the Ides were reckoned as days *before* the Calends of the succeeding month. This system was still in use in western Europe as late as the sixteenth century![59]

In Imperial times, however, the custom became popular, under astrological influence, to use the seven-day week with the different days named after the respective 'planets'.[60] Inscriptions at Pompeii list the 'days of the gods', namely Saturn, the Sun, the Moon, Mars, Mercury, Jupiter, and Venus. This order, from which our modern days of the week derive (e.g. in French), appears at first sight to be devoid of sense, since it does not accord in an obvious way with the order in which (according to pre-Copernican cosmology) the 'planets' were thought to lie in relation to the earth: Saturn, Jupiter, Mars, the sun, Venus, Mercury, the moon. The explanation is that the planets were believed to rule the hours of the day as well as the days of the week and that each day was associated with the planet that rules its first hour. The first hour of Saturday was ruled by Saturn, and similarly the eighth, fifteenth, and twenty-second hours. The twenty-third was allotted to Jupiter, the twenty-fourth to Mars, and the first of the next day to the sun, which thus ruled Sunday, and so on through the week. From this comes the custom, introduced in the third century AD, of indicating the most important dates according to the weekdays as well.[61]

The Christians, because of the Jewish origin of their religion, at first adhered to the Jewish seven-day week in which the days, except the Sabbath, were numbered but not named. In due course, however, they began to be influenced by the astrological beliefs of converts from paganism; as a result, they adopted the planetary week. Meanwhile, the influence of Mithraism had led the pagans to substitute *Dies Solis* (the Sun-day) for the *Dies Saturnis* (the Saturn-day) as the first day of the week. This change appealed to the Christians who had long observed Sunday—the Lord's Day (*Dies Dominica*) on which Christ rose from the dead—as the first day of the week, in place of the Jewish Sabbath. The planetary week was officially adopted in AD 321 by the Emperor Constantine, who also followed the Christian practice of regarding Sunday, instead of Saturday, as the first day of the week. He formally decreed that magistrates, citizens, and artisans were to rest from their labours 'on the venerable day of the sun', but he permitted field work. Already in the first century AD, under the influence of Judaism, Roman Society had begun to introduce a weekly day of rest—unlike ancient Greece where there were not even any school holidays, except on special occasions such as days in honour of Apollo, Poseidon, etc.[62] Tertullian (*c*.155–222) was the first Church Father to declare that Christians ought to abstain on Sunday from secular duties or occupations, lest these should give pleasure to the Devil.

The first mention of Christmas Day, as far as we know, was in the Roman calendar for the year 354. Previously, 6 January had been celebrated as the Epiphany, or anniversary of Christ's baptism, which was believed to have occurred on his thirtieth birthday. The choice of 6 January for this purpose has been traced back to the gnostic Christians of Egypt, the corresponding date in the calendar used there being traditionally associated with the blessing of the Nile. Christ's birthday only became important for the Church when infant baptism replaced adult baptism. This led to the belief that Christ's divine nature originated at his birth rather than at his baptism. As a result, by about the year 400 Christmas Day had become a significant date in the Christian Year: 25 December was chosen so as to exorcize the great pagan festival of the solar solstice.

In the latter part of the fourth century the last great emperor of the west, Theodosius, who was of Spanish origin, finally abolished the pagan Roman calendar with its hotchpotch of festivals, thereby severing one of the most familiar links the Romans had with their historic past. Consequently, it is to him that the European world owes a uniform calendar corresponding to the needs of a universal society and based upon the Christian year. In 386 he reaffirmed his decree and invoked severe sanctions against those who desecrated the Lord's Day.[63] The view that the Lord's Day is essentially the Jewish Sabbath—a 'taboo' day—transferred from the seventh to the first day of the week found expression from time to time in medieval law and theology. It culminated in the sabbatarian excesses of English and Scottish Puritanism and the Sunday legislation, much of which has been relaxed since the First World War.

Easter was introduced in Rome about the year 160, and as in Alexandria was celebrated on the Sunday following the Hebrew Passover, which for practical purposes could be reckoned as the Sunday following the first full moon after the spring equinox. A set of Easter tables drawn up by Cyril of Alexandria (376–444) was accompanied by a consecutive set of years beginning with the Emperor Diocletian and his persecution in AD 284, but when in AD 525 a Scythian monk living in Rome, Dionysius Exiguus, prepared a continuation of Cyril's tables, at the request of Pope John I, he felt that it was inappropriate to reckon from the reign of this enemy of Christianity, and he chose instead to date the years from Christ's Incarnation.[64] Astronomical evidence suggests that this may have occurred in the first half of the year 5 BC. (For astronomers, unlike historians and chronologists, there is a year 0.) Although Dionysius' system was the origin of the AD sequence that we now employ, it was not made use of for nearly 200 years, the oldest known work in which it is employed being Bede's *Ecclesiastical History of the English Nation*, of the early eighth century. The BC system, extending backwards from the birth of Christ, was occasionally used by Bede, but after him it lapsed until the fifteenth century. It did not come into general use until the latter half of the seventeenth century.

5. Time in the Middle Ages

Medieval Europe

In the year 430 the Vandals, who had crossed the Mediterranean not long before, were battering at the walls of St Augustine's home town as he lay there dying. This can be regarded as symbolic. For, in its great days, particularly in the age of the Antonines in the second century, the Roman empire had been primarily a civilization of towns. These were very different from the haphazard constructions of medieval Europe. They were deliberately planned with their streets laid out in orthogonal grid-systems like the great cities of Hellenistic times, such as Alexandria and Antioch. The decay of the Roman empire was most clearly revealed by the decline of towns and increasing ruralization. This transformation occurred primarily in the northern and western provinces, which were always more of a liability and less a source of wealth and culture than the southern and eastern provinces. For example, Africa supplied Rome with two-thirds of its corn, transported in the great grain ships which must have been among the most impressive sights of antiquity. The northern and western provinces were comparatively much less developed and the principal towns in them, such as Segovia, Arles, York, and Cologne, were primarily military camps.

Among the causes of the fall of the Roman empire were successive attacks by barbarians. Although in the sixth century the Byzantine Emperor Justinian's great generals Belisarius and Narses succeeded in reconquering much of the west, so that for a time the Mediterranean again became a Roman lake, in the following century Europe faced a dangerous new enemy. Fanatical warriors inspired by a new and militaristic religion, Islam, brought about the final break between East and West. By the year 700 learning in western Europe was confined to Ireland and the coast of Northumbria. The only centres of learning were the monasteries in those remote areas, and it is in one of these, founded in 682 at Jarrow by a wealthy Northumbrian nobleman turned monk, Benedict Biscop, that we find 'the first scientific intellect produced by the Germanic peoples of Europe'.[1] The Venerable Bede (673–735) spent most of his life at Jarrow as a Benedictine monk, praying, reading, and teaching Latin, Greek, and Hebrew. He was ordained priest by St John of Beverley, so gaining the title 'Venerable', which was a rare dignity among monks, but the usual form of address for a priest at that time. A really great scholar, he had a unique opportunity to develop his abilities, because Biscop had brought back to Jarrow some 200 to 300 antique books that he had managed to acquire in southern Italy. Bede also had access to the library collected by Bishop Acca at Hexham.[2] He was thus able to acquire an unusually extensive knowledge for his day of ancient literature, including the works of St Augustine and the scientific writings of the elder Pliny.

Bede's main object in life was to transmit his knowledge in intelligible form to his contemporaries and successors, and in this he was eminently successful. As Sir Arthur Bryant has so vividly put it, 'That life of scholarship and labour, with the tireless hand writing amid the intervals of prayer and teaching, sometimes so frozen that it could hardly grip the pen, is one of the proud memories of England.'[3] In all Bede wrote thirty-five works, of which twenty were commentaries on Exodus, Proverbs, and other books of the Bible and six were works of chronology. His most famous book *The Ecclesiastical History of the English Nation* was the first historical work produced in England. Written in Latin, it was translated into English towards the end of the ninth century by Alfred the Great. Containing a greater proportion of secular matter than the Church history of Eusebius, it was based partly on written material and partly on the memories of men still living. A considerable part of medieval historiography was based upon it. In particular, Bede had a direct influence on the Carolingian renaissance of the ninth century through his pupil Egbert, who became Archbishop of York and trained Alcuin, who under Charlemagne founded the Frankish schools that did so much to stimulate learning on the Continent.

Bede's writings are of considerable importance in the history of chronology. This had already become a subject of crucial significance in England during the course of the seventh century. Although the death in battle of Penda, the pagan king of Mercia, in 655 sealed the victory of Christianity over heathenism, this important event was overshadowed by dissension between the Roman and Celtic Churches. The principal source of discord concerned the date of Easter. Our present rules for its determination (see Appendix 3), as set out in the *Book of Common Prayer*, follow the Roman tradition whereby Easter Day is the first Sunday after the first full moon following (or on) 21 March. But if full moon occurs on a Sunday then Easter Day is the following Sunday. The reason for this was to avoid concurrence with the Jewish Passover. (The expression 'full moon' means the fourteenth day of the moon reckoned from its first appearance.) The Celtic Church, founded in the sixth century by St Columba with the aid of Irish-trained monks, followed Rome in always celebrating Easter on a Sunday, unlike the eastern Churches; but owing to its remoteness the Celtic Church experienced difficulty in being kept fully informed of doctrinal and other changes decided on in Rome. Consequently, unlike Canterbury, it failed to keep in line with the Roman practice when the fourteenth day of the moon fell on a Sunday. As a result, by the middle of the seventh century a peculiar difficulty had arisen in Northumbria. For, although King Oswy followed the Celtic practice, his consort, Queen Eanfleda, who had with her a Kentish priest named Romanus, adhered to the Roman practice. Most years this gave rise to no special problem, but eventually there was an occasion when the king's enjoyment of the Easter Feast was spoilt by the absence of his queen, who was still fasting because for her it was Palm Sunday.

To resolve the problem of Easter and other points of dispute between the Churches, Oswy convened the Synod of Whitby in 664. In chapter 24 of his *Ecclesiastical History* Bede gives an account of what took place. Oswy was probably unable to follow in detail the abstruse arguments put forward, but in the end he decided to accept the Roman practice, on the grounds that at the gates of heaven the keys are held by St Peter, and against him he would not contend. 'The king having said this, all present, both great and small, gave their assent, and renouncing the more imperfect institution resolved to conform to that which they found to be better.'[4] Henceforth, the English Churches were to have the advantages of the unity and discipline that the Church of Rome had inherited from the Empire.

Bede not only compiled a detailed account of this important Synod but in another of his treatises, *De temporum ratione* ('*On the Reckoning of Time*'), written in 725 and generally regarded as his scientific master-piece, he computed Easter tables for the period 532 to 1063 and also made a first attempt at a general chronology of the world down to the reign of the contemporary Byzantine emperor, Leo the Isaurian. Chapter 29 of that work is remarkable for containing the first scientific investigation of the tides, involving the earliest 'establishment of a port', that is, the mean interval between the time of high water and that of the previous transit of the meridian by the moon.

It was also through Bede that the AD system of reckoning the years from the Incarnation of Christ, that had been devised two centuries before by Dionysius Exiguus, was introduced into England. Dionysius' cycle of the year began with 25 March, the Annunciation of the Blessed Virgin Mary. From the time of Bede the Christian era became established for the dating of charters, but at first only in England. According to R. L. Poole, 'It passed to the Continent by the means of Anglo-Saxon missionaries and scholars. St Boniface took it with him into the Frankish kingdom. But it does not appear to have been regularly employed in the Royal Chancery until the last quarter of the ninth century, from which time it became a fixed element in diplomas.'[5] It was not until the pontificate of Pope John XIII, elected in 965, that the Papacy began dating by the year of the Incarnation, but the practice was not uniformly adopted until the time of Pope Leo IX, elected in 1048.

Chapter 35 of Bede's *De temporum ratione* is the *locus classicus* of the concept of the 'ages of man', the medieval division of human life into a number of distinct periods best known to us today through the speech on the 'seven ages of man' by Jaques in Shakespeare's *As You Like It* (Act II, scene 7). Most ancient and medieval writers thought of human life not as a continuous development but instead as punctuated by a number of sudden changes from one 'age' to the next. (This idea was extended to prehistory by the social anthropologist A. van Gennep who, in 1909, introduced the term *les rites de passage* for the rituals originally associated with such changes in the life of the individual.) Bede was the first Englishman to describe the theory of the *four* 'ages of man'. For the source of this we must go back to the Pythagoreans of the sixth century BC, whose cosmological speculations were based on the '*tetracys*', that is, the geometrical symbol composed of ten discrete points symmetrically arranged in the form of an equilateral triangle with sides of four points each. The number four came to be associated with many natural phenomena, for example the four seasons, the four cardinal directions, and the four elements of the Greek theory of matter from Empedocles to Aristotle.

For some 2,000 years great significance continued to be attributed to the number four. For example, long after Bede, in his *Boke of Nurture* John Russell, who had been Marshal of the household of the great patron of learning, the youngest son of Henry IV, Duke Humphrey of Gloucester (1391–1447), described how the four courses of an elaborate fish-dinner that he had prepared for his master and guests was accompanied by appropriate 'subtleties', or ornamental devices. During the first course Duke Humphrey's guests were to contemplate the representation of a 'galaunt yonge man' standing on a cloud (signifying the element 'air') at the beginning of spring (associated with the sanguine humour). During the next course they were faced by the representation of a 'man of warre' standing in fire (associated with summer and the choleric humour); and while consuming the third course they were confronted by the form of a man with a 'sikelle in his hande' standing in a river (signifying water and the phlegmatic humour associated with autumn and harvest-time). The fourth and final course, which came with spices and wine, ushered in a representation of winter in the form of a man 'with his lokkys grey, febille and old' sitting on a cold hard stone (signifying the element 'earth' and the melancholy humour). 'Thus,' as J. A. Burrow remarks, 'as Duke Humphrey's guests worked their way through this very unpenitential fish banquet, they were invited to see in it the four courses of their own life's feast.'[6]

Although Bede discusses the theory of the 'four ages' and even refers to the alternative concept of the 'six ages', he makes no mention of the 'seven ages' later described by Shakespeare. He could not mention it because it was not known in the Latin West before the revival of learning in the 'renaissance of the twelfth century' (a useful term due to the American medievalist Charles Homer Haskins, who introduced it in 1927). The idea of the 'seven ages', unlike that of the four, was astrological in origin. It goes back to the astronomer Ptolemy of Alexandria (*fl. c.* AD 150), seven being the number of the 'planets', including the sun and moon. This idea is fully described in Ptolemy's *Tetrabiblos* (iv. 10). (A translation into English, by F. E. Robbins, was published in 1940. An excerpt appears on pp. 197–8 of the Appendix to the book by Burrow cited above.)

Although through the efforts of Charlemagne, who was crowned Holy Roman Emperor by the Pope in the year 800, the centre of European culture began to move northwards from the Mediterranean, the Viking raids of the ninth and tenth centuries delayed the full effects of this until about the year 1000. England particularly suffered from these raids, so that by the time of the twelfth-century renaissance it was, in the words of R. W. Southern, 'a colony of the French intellectual empire, important in its way and quite productive, but still subordinate'.[7] The main creative activity of the English monastic houses was in historiography. With the notable exception of Bede and the authors of *The Anglo-Saxon Chronicle*, earlier generations had not, on the whole, been greatly interested in historical records, but the Conquest brought about a great transformation. The Normans insisted on the production of titles to estates and threatened to confiscate those for which none were forth-coming. In these conditions, the Conquest convinced the English monasteries that corporate survival depended on the discovery and preservation of the past. Consequently, as Southern has argued, 'history was not simply an adornment: it was a necessity'.[8]

One of the principal features of the renaissance of the twelfth century was a great increase in historiography, stimulated not only by the Norman conquest of England but also by the crusades and the rise of the north Italian communes, or city-states. Moreover, men such as the great ecclesiastical architect Abbot Suger of St Denis (*c.*1081–1151), who devoted his later years to composing a laudatory life of the French monarch Louis VI (1081–1137), effectively a second founder of the Capetian dynasty, wrote history with the object of producing favourable propaganda rather than documentary facts. Suger's historical writings led the monks of his abbey to develop a taste for history and hence to compile a series of chronicles. In the same century universal history also flourished, mainly with the object of determining the end of the world, after the year 1000 had passed without any sign of its impending occurrence. This type of history was, of course, theologically rather than politically orientated. The influence of the twelfth-century apocalyptic historians was destined, however, to be soon overshadowed by that of Joachim of Fiore (see pp. 81–2).

Among the technical arts cultivated in some continental schools that began to affect England soon after the Norman Conquest were those of measurement and calculation. Haskins has drawn attention to interest-ing evidence for this in an autobiographical fragment written by a Benedictine prior, Walcher of Malvern, whose tomb survives there bearing the date 1125. In it he refers to the lunar eclipse of 30 October 1091 that he happened to observe in Italy. On returning to England, he discovered that several hours appeared to have separated the time of the eclipse in Italy and in England. Puzzled by this, he was careful later to record the time as precisely as he could when, unexpectedly, on 18 October of the following year the moon underwent another eclipse:

I at once seized my astrolabe and made a careful note of the time of full eclipse, which was a little more than three-quarters of an hour after the eleventh hour of the night. If this time is converted into equinoctial time, it will be found to be shortly before 12.45. Hence, according to this rule which I have explained earlier, the lunar cycle began on 3 October at 19.30 hours.[9]

As Southern has remarked, this passage, only part of which I have quoted, illustrates the difficulties encountered in those days of telling the time and Walcher's anxiety for precision in seeking to establish the exact correlation between the phases of the moon and the solar calendar.

To men of the Middle Ages astronomy was of particular interest because it seemed to offer the best means of understanding, and possibly controlling, terrestrial events. An essential tool for enabling astronomers to advance beyond the stage reached by Bede was the astrolabe. This instrument had been introduced in the West in the eleventh century from the world of Islam, which in those days enjoyed a higher degree of civilization and of scientific and technological expertise than the West. For anyone in northern Europe to attain a proper understanding of Islamic science it was necessary to go abroad. Among the first to do so for this purpose was Adelard of Bath (*fl.* 1116–42). He first went to Paris, but not finding what he wanted there, he moved on to Salerno in southern Italy and then to Sicily, where he learned Arabic. Later he probably visited Spain. His outstanding role in the development of science in the Latin West was due to his translations from the Arabic, which were of a crucial and seminal nature.

The Islamic world

The origin of Islamic interest in science can be traced back to the closure by Justinian of the Neoplatonic Academy at Athens in 529. Scholars from there were invited to Iran, and they brought much Greek learning with them. Interest in the subject having thus been aroused among learned men in western Asia, a scientific institute was eventually set up in Baghdad after the Muslim conquest of much of that region. It attained its highest reputation during the caliphate of al-Ma'mun (813–33), son of Harun-al-Rashid of *Arabian Nights* fame, and himself an astronomer. By the end of the ninth century many Hellenistic scientific and technological works had been translated into Arabic, including Ptolemy's great astronomical book *Syntaxis*, which is usually known today by its Arabic title *The Almagest*. As a result of all this activity, Baghdad was the true successor of Alexandria, the former intellectual capital of the Hellenistic world. Knowledge of Greek science and technology, combined with Iranian and Indian traditions and enhanced by further scientific studies and inventions, spread from there to other parts of the Islamic world, including Sicily and southern Italy and especially Moorish Spain, where by the twelfth century the main centres of learning were in Cordoba and Toledo.

Muslims in all parts of the Islamic world required mathematically educated persons who would be able to determine the astronomically defined times of prayer and the direction of Mecca. It is, therefore, not surprising that many portable instruments for the determination of time were required, including the chief instrument used by both Arabic and Latin astronomers, the astrolabe. This instrument was known to Ptolemy in the second century AD, and the underlying mathematical theory of stereographic projection can be traced back at least to Ptolemy's great predecessor Hipparchus (second century BC).

The form of astrolabe used in medieval Europe, however, was derived from the Muslim type found in Spain. A good English account of it was given by the poet Chaucer in the second half of the fourteenth century. It consisted of a circular metal plate (usually brass) graduated in degrees around its rim. It was marked with a datum line (or diameter) and hinged to its centre was a rotating line (or pointer). Portable models could be hung from a ring on the rim so that the datum line was horizontal. By directing the pointer at a particular star, its altitude could be read off against the scale on the rim to an accuracy of about one degree. For any given latitude the Pole star has effectively a constant altitude and the other stars appear to revolve around it owing to the Earth's diurnal rotation. On the front of the astrolabe there was a thin plate (the tympan) on which was engraved a stereographic projection of the lines of altitude and azimuth (angular distance along the horizon) as they would be for an observer at a given latitude. An open-work star map in stereographic projection (known as the rete) was in front of the tympan, and could be rotated by hand over the lines of altitude and azimuth.

An early form of analogue computer, the astrolabe was primarily designed to solve problems of spherical trigonometry to shorten astronomical calculations.[10] From the scales engraved on it, it was possible to determine the positions of the so-called 'fixed stars' in relation to the horizon and of the sun, moon, and planets in relation to the stars. Designed for the latitude of a particular place, its most important use was to determine the precise time of day or night from an observation of the altitude of the sun or one of the stars mapped on the rete, but of course by modern standards the result was not very accurate. Moreover, although the astrolabe enabled long calculations to be avoided, the computing of planetary positions, for example, for casting a horoscope, still involved a considerable amount of work.

As regards other time-measuring instruments, extensive remains of two monumental Islamic water-clocks still survive at Fez in Morocco.[11]

A book in Arabic *On the Construction of Water-clocks*, believed to be partly based on the translation of a Hellenistic treatise in Greek, preserves the idea of the invention of the basic machinery of a water-clock by Archimedes, together with later ingenious additions to the mechanism made by either Byzantine or Islamic craftsmen. It was probably composed after 1150. It has recently been edited and translated into English by D. R. Hill, who points out that 'horologically, it provides an important link between the water-clocks of the Hellenistic world and those of Islam'.[12] Detailed discussion of some other Islamic clocks will be found in a book written in Baghdad about 850 and also translated by D. R. Hill.[13]

One special case in which the influence of Islam made an important cultural contribution to the development of temporal concepts in Europe concerns music. Early medieval church music was all plain chant, in which the notes had fluid time values. Mensural music, in which the durations of the notes had an exact ratio among themselves, seems to have been an Islamic invention. It was introduced into Europe about the twelfth century. It was at this time too that there appeared in Europe the system of notation in which the exact time-value of a note is indicated by a lozenge on a pole.

As regards the theoretical and philosophical analysis of time, the most important and original contribution of medieval Islamic thinkers was their theory of discontinuous, or atomistic, time.[14] The most famous exponent of this concept, but not its originator, was the twelfth-century philosopher Moses Maimonides, who wrote in Arabic although he was a believing Jew. In the most celebrated of his works, *The Guide for the Perplexed*, he said: 'Time is composed of time-atoms, i.e. of many parts, which on account of their short duration cannot be divided. . . . An hour is, e.g. divided into sixty minutes, the second into sixty parts and so on; at last after ten or more successive divisions by sixty, time-elements are obtained which are not subjected to division, and in fact are indivisible.'[15] This atomistic view of time was associated with a drastically contingent and acausal concept of the world, its existence at one instant not implying its existence at any subsequent instant.

D. B. MacDonald has speculated on the difficult question of the origin of this view in Islam and has suggested that it arose from a Muslim heresy 'in that dark but intense period of theological and intellectual development which stretched from the death of Muhammad for at least two and a half centuries'.[16] The atomistic theory of Epicurus, the methods of the Greek sceptics, and Zeno's paradoxes concerning time and space may all have influenced the heretics concerned, but MacDonald could find no trace of any Greek theory combining material and temporal atomism and sought instead to attribute the occurrence of the latter in Islamic thought to Indian influence.

The Islamic calendar is one of the few remaining purely lunar calendars, the year being just over ten days shorter than the tropical year, or year of the seasons. The Islamic era began on 16 July 622, the first day of Muhammad's flight to Medina. The circumstances in which this was adopted as an epoch, instead of the time when the Prophet was either born or entrusted with his divine mission or died, are explained by al-Biruni (AD 973–*c.*1050) in his great work *The Chronology of Ancient Nations*.[17] The fundamental instant in Islamic life occurs with the new moon, which must be watched for and established by two 'witnesses of the instant'.[18] The 'perfect instant', however, is the Hour of the Last Judgement, for the 'witness' of this instant is the divine Judge himself.

The periodization of history and millenarianism

This type of eschatological view of time was, of course, not confined to Islam, for we find it also in Zoroastrianism, Judaism, and early and medieval Christianity. In the Christian case, it led to the periodization of history, a chronological method that we still use, although nowadays we approach history from a purely secular point of view. Medieval historians followed the scheme devised by St Augustine for dividing world history into six ages corresponding to the six days of Creation described at the beginning of Genesis. Living in the troubled times of the late fourth and early fifth centuries, St Augustine regarded the Christian era as the age of senility and decay that would lead to the seventh age when time would end, although he was careful not to forecast a definite date for this. The most important change in the Christian outlook on history between the Apostolic age and that of St Augustine was the gradual realization that the end of the world was not at hand. He laid particular emphasis on those passages in the New Testament (e.g. Mark 13: 32) that emphasize our total ignorance of when the Second Coming will occur. Bede too believed that the time of Doomsday is concealed from mankind.

As Beryl Smalley has said, 'The concept of the six ages saddled medieval historiographers with a gloomy picture of their times.'[19] But, even though it discouraged optimism and ruled out the possibility of progress, it did not weigh too heavily on medieval historians, particularly because the year 1000, which had been awaited by many with a mixture of hope and trepidation, had passed without any sign of the world coming to an end. Many prophets in the tenth century believed that the world would come to an end in the year 1000, but according to A. J. Gurevich the legends concerning mass psychoses in Europe as the year 1000 approached originated at the end of the fifteenth century when people really were afraid that the end of the world was imminent.[20]

Millenarian belief arose from combining the idea expressed in Psalm 89: 4 that 'A day with the Lord is as a thousand years' with the interpretation of the Sabbath, or seventh day, as a symbol of heavenly rest in accordance with Hebrews 4: 4–9. The most influential exponent of millenarian beliefs in the Middle Ages was Joachim of Fiore (1145–1202). He was a Cistercian monk who became Abbot of Curazzo in Calabria, in southern Italy. This was a part of the world where Greek culture and the Roman Church met and where there was a strong Saracen influence. Consequently, it was a region subject to many cross-currents of thought and belief. Joachim eventually broke away from the Cistercians and retired to a lonely spot in Calabria where disciples gathered around him and he was given papal permission to found his own congregation. Joachim's idea of a completely unworldly religious order nearly found expression in the confraternity that began to form around the followers of St Francis of Assisi shortly after Joachim's death, but the main body of Franciscans soon made concessions to the demands of everyday life.[21] Eventually, in 1570, Joachim's community was absorbed by the Cistercian order.

Joachim called his monastery San Giovanni in Fiore in expectancy of the new life that must come to flower.[22] He was a keen student of the scriptures, particularly the Book of Revelations, and while meditating in his Calabrian retreat on the mystery of the Trinity and how it related to the time-process he had moments of intense spiritual illumination that led him to formulate a new millenarian philosophy of history. He laid great emphasis on the unity of the Trinity, arguing that root, stem, and bark together form one tree. Joachim claimed that there are, however, three distinct ages or states: that of God and the Old Testament, which was the age of fear and servitude; that of Christ and the New Testament, which is the age of faith and submission; and the Third Age of the Ever-lasting Gospel, or Age of the Holy Spirit, that will supersede the Old and New Testaments and be the age of love, joy, and freedom. His fervently expressed hope in the coming of the Age of the Holy Spirit may have had its origin in the Jewish concept of the Messianic Age, for like the latter it was regarded by him as lying essentially within history and not beyond it, being indeed the climax of history. This belief was totally irreconcilable with the Augustinian view that, in so far as it is possible for the Kingdom of Heaven to occur here on earth, it has already been realized in the Church.[23] Joachim's concept of history was far more dynamic than St Augustine's. As an authority on the influence of Joachimism has remarked:

'What characterizes this Christian revolutionary tradition from Joachim of Fiore to John Huss, from Thomas Münzer to the theologies of hope and political theologies of our own day, is that the Kingdom of God is not conceived as another world in space and time, but as a different world, a changed world, a world changed by our own efforts. . . . This means that human history is where all the issues are settled.'[24]

Joachim had a profound influence on later prophecies down to the end of the seventeenth century. It is difficult for us now to understand how it was that so many serious thinkers in those days were prophetically-minded. Even Isaac Newton (1642–1727), although not directly influenced by Joachim, devoted much of his time to the correlation of prophecy, history, and the end of the world.[25] *Granted his initial assumptions*, however, he was in fact just as scientific in his calculations in that field as in his famous contributions to mathematical physics and astronomy.

The measurement of time

In his well-known book *Feudal Society*, the historian Marc Bloch has laid particular emphasis on the fact that in the Middle Ages men found it difficult to appreciate the significance of time because they were so ill-equipped to measure it. For not only were water-clocks rare and costly, but in countries such as England, northern France, the Netherlands, and Germany sundials were inadequate because skies were so often cloudy. According to Asser's *Life of King Alfred*, that intellectual monarch had candles of equal length lit successively to mark the passing of the hours; but, as Bloch remarks, 'such concern for uniformity in the division of the day was exceptional in that age.'[26] To illustrate the point, he describes an incident recorded in a chronicle of Hainault concerning a judicial duel that was to take place at dawn. Only one contestant appeared and at the end of the prescribed waiting period, the hour of nine he asked for the non-appearance of his adversary to be legally recorded. The judges had to decide whether the time limit had been reached. They deliberated, looked at the sun and then questioned the clerics, since the practice of the liturgy and the regular tolling of church bells had accustomed them to a more precise knowledge of the rhythm of the hours than the judges themselves possessed. As Bloch has commented, 'To us accustomed to live with our eyes constantly turning to the clock, how remote from our civilization seems this society in which a court of law could not ascertain the time of day without discussion and inquiry!'[27]

One of the peculiarities revealed in many surviving documents from the Middle Ages is the lack of precision with which the times of events and measurements of duration were recorded. John Nef, in his Wiles Lectures of 1956, concluded that, if we seek the origins of our modern quantitative-mindedness, we must concentrate on the last decades of the sixteenth century.[28] Earlier we find little trace of it generally, and so we ought not to be surprised to find it missing from the ordinary person's consciousness of time in those days. In his book *Time in French Life and Thought* Richard Glasser has drawn attention to the fact that nowhere in the *Chanson de Roland* do we find any indication of time. The epic poet 'was aware neither of the falling of leaves in autumn nor of the passing away of generations. These were phenomena which in no way attracted his attention. The essential quality of the world was its transitoriness *vis-à-vis* God, not the visible change which went on unceasingly in the world.'[29] Until the fourteenth century only the Church was interested in temporal measurement and division. Even the concept of the hour was not used as a unit of duration before the time of Middle French. In the popular tongue it was used only to indicate a point in time.[30]

In view of the slowness with which changes of mental outlook came about in those days, it is not surprising that even after the introduction of the mechanical clock in the fourteenth century most people, including many of the more sophisticated, were far less concerned in their daily life with the passage of time than we are. A striking example is provided by a famous maker of astronomical instruments, Jean Fusoris, who was arrested on suspicion of treason in 1415, during the invasion of France by Henry V. Interrogated twice in a single year, on the first occasion he claimed to be 'fifty or thereabouts' and on the second 'sixty or thereabouts'![31]

In England, parish registers providing dates of birth were instituted by law in 1538. Previously when someone's age had to be formally determined it had to be done in the presence of a sheriff of the county and a 'jury' composed of local people who knew the person concerned. This procedure was followed when a minor inheriting property claimed to have become of age, or when it was thought necessary to determine legally that someone had attained the age when he or she was allowed to marry. Of course, the indifference to time generally attributed to medieval people was not absolute. Already by the year 1200 there were numerous signs of economic pressure on time, and even two centuries earlier it appears that peasants and artisans near Fleury tended to ignore feast-days through a need to work in their fields.[32]

Another indication that our medieval forebears had very different standards from ours for recording the lapse of time is revealed by the way in which they dated their letters. As late as the fifteenth century it is doubtful whether people in general knew the current year of the Christian era, since that depended on an ecclesiastical computation and was not used much in everyday life. They seldom dated their letters and when they did it was by the year of the king's reign. Even when chroniclers of the period gave the year of our Lord it was often wrongly stated. This is not surprising since different numbers were assigned to the year in different places. R. L. Poole has given the following hypothetical example to illustrate this:

If we suppose a traveller to set out from Venice on March 1, 1245, the first day of the Venetian year, he would find himself in 1244 when he reached Florence; and if after a short stay he went on to Pisa, the year 1246 would already have begun there. Continuing his journey westward he would find himself again in 1245 when he entered Provence and on arriving in France before Easter (April 16) he would be once more in 1244.[33]

This seems a bewildering tangle of dates, but as a rule the traveller would take note only of the month and the day. If, however, he did consider the year it would be that of the place where he usually lived. In practice, only writers of documents and chronicles were concerned with the number of the year.

Months and days were, of course, more likely to be correctly stated, and letters were frequently dated in this respect but much more use was made of festivals and saints' days. In his Introduction to the *Paston Letters* J. Gairdner pointed out that letters were often dated as being written on a particular day of the week, say Monday or Wednesday, *before* or *after* such a celebration. For example, Agnes Paston even dated a particular letter (No. 25) during the week by reference to the Collect of the previous Sunday: 'Written at Paston in haste, the Wednesday next after *Deus qui errantibus*.'[34] The modern practice of numbering the days of the month consecutively from the first to the last came to the West from

Syria and Egypt in the second half of the sixth century. Pope Gregory VII introduced it into his chancery, but his successors reverted to the old Roman style. The revival of learning under Charlemagne (*c.*800) was in the Latin tradition, and so there was an official reversion to the Roman style in the Imperial chancery too which persisted for centuries.

A far more modern attitude to time and dates was adopted in the previous century by the famous Italian poet and reviver of classical literature Petrarch (1304–74). Time was the theme that fired his heart as a young student and affected him for the rest of his life. Because he kept a detailed record of the temporal milestones in his life we have more precise information about him than of anyone who lived before him. In all his writings, poetry as well as prose, he maintained what has been described as 'an attention nothing less than astounding to exactitude in date'.[35] Moreover, unlike most medieval letter writers—and, for that matter, even unlike most of us today when we dash off our epistles without much thought about the time—Petrarch 'spells out the dates (including the hour) with weight and deliberation, as if to stress the importance of taking one's bearings in time'.[36] For example, in a letter written in 1364 he was careful to give the precise hour of arrival of the boat that brought news of the Venetian victory against Crete. 'It was, I believe, the sixth hour of June 4, this year 1364.' Although time seems always to have been important for Petrarch, he tended to value it even more as he got older because he realized that, as with other things, it becomes more precious as it becomes less plentiful. In his responsiveness to temporal processes he differed from many of his contemporaries and we can look upon him as the forerunner in literature of those, like Spenser and Shakespeare in the late sixteenth century, who were greatly concerned with the irreversible effects of time on the human mind and spirit. Although western-European society in the Middle Ages developed no general concept of progress, many important innovations were made. Indeed, in technology western Europe advanced far beyond the Roman empire. The Romans were in some respects good engineers, as is evident from their sophisticated heating systems involving plumbed hot water and their networks of roads, but in other ways they were often surprisingly primitive. Apart from those transmitted from China, medieval inventions included, for example, spectacles for reading, the spinning wheel, stronger iron tools than had been previously available, the heavy plough, and the use of coal as a fuel. Moreover, in the building of the great Gothic cathedrals many new devices were introduced, including flying buttresses. Some of the most important innovations in the Middle Ages were connected with the use of the horse as a source of motive power. A more efficient harness than the crude yoke, which had been so well suited for draught-oxen, was introduced about the ninth century. Late that century Alfred the Great noted, with apparent surprise, that horses were used for ploughing in Norway.[37] This would have been impossible with the yoke-harness, because as soon as the horse begins to pull with it the neck-strap presses on the animal's windpipe and thus tends not only to restrict the flow of blood to its head, but also to suffocate it!

Another important development was the iron horseshoe that was nailed on to the hoof. Previously, the shoe had only been tied on and this greatly impeded the animal's progress. The first indisputable evidence of the use of nailed horseshoes goes back to the ninth century. The development and the elaboration of metal armour for protection in warfare and jousting gave considerable impetus to the craft of the blacksmith. This was destined to be of particular importance to the measurement of time, because the blacksmith was the forerunner of those who constructed the first mechanical clock. It is surely significant that one of the greatest of these, Richard of Wallingford, Abbot of St Albans in the early fourteenth century (see ch. 7), was the son of a blacksmith.

6. Time in the Far East and Mesoamerica

India

It has already been suggested (ch. 5), that the Islamic atomistic theory of time may have been the result of Indian influence. In discussing this possibility, MacDonald has drawn attention to an article on 'Atomic Theory (Indian)' by Hermann Jacobi.[1] In it Jacobi referred to the theory of the momentariness of all things formulated by the Sautrânkitas, a Buddhist sect which originated in the second or first century BC. According to that theory everything exists for only an instant and is then replaced by a facsimile of itself, so that it is but a series of momentary existences like the successive frames in a cine-camera film. The concept of entities that appear for only an instant and then disappear was used by Buddhists to prove that all is merely appearance and that absolute reality does not fall within the domain of the intellect. But how and why this atomistic temporal concept, which Buddhism used for its own purposes, was adapted to the very different objects of Islam remains an open question.

In Classical antiquity there were connections between Europe and India even before the conquests of Alexander had extended as far as the north-western part of the Indian subcontinent. Already in the sixth century BC, at the time when Buddha and Mahavira lived, that part of India was ruled by the Achaemenids of Iran, and Iranian influences have been important in India ever since. As in the Zurvanite system, philo-sophical speculations concerning time formed the essential part of a particular Indian philosophy known as the Kalavada that was later absorbed by other systems. The term *kala* was originally employed by the Hindus in the Rig-Veda to denote the 'right moment' in connection with sacrificial ritual. Later it came to denote 'time' generally, and it was usually employed in that sense in Sanskrit writings. In the Vedic period the abstract idea of time was regarded as the fundamental principle of the universe, but whether it was made into a deity is uncertain. The word *kala* has been associated, however, with Kali, 'the Black One', one of the forms of the consort of the god Siva. Time was regarded as black and connected with Siva, god of destruction, because it is hard and pitiless.

Anindita Balslev has recently drawn attention to the subtlety of many of the Hindu philosophical arguments, for example that concerning the perceptibility of time which took place in the eleventh century. On the one hand, the Bhatta-Mimamsaka school argued that time is perceptible, whereas their Nyaya-Vaiseka opponents claimed that it is only an inferred concept because it lacks sensible qualities, such as colour, form, etc. The former school maintained that sensible qualities are not the sole criteria of perceptibility and that time is perceived always as a qualification of sensible objects. In other words, events are perceived as quick, slow, etc., and these properties involve a direct reference to time. Their opponents retorted that time *per se* cannot be perceived and that inference is the only means of our knowing time as an ontological reality.[2] Other subtle philosophical discussions concerned the contrast between the objective reality of the instant and the ideal nature of duration, because the latter is a mental construct, whereas the former is experienced (the opposite of what we think in the West today).

Indians wrote no historical books with numerical dates and regarded personal life as one of a succession of lives of the same individual repeated infinitely often in endless time. This idea of metempsychosis, or transmigration of souls, has only occasionally appeared in the West, in particular in the school of Pythagoras, which may have been subject to Eastern influences, since he was roughly contemporaneous with Buddha—and also with Zarathustra. Although a minor error in the recitation of the Vedas was looked upon with strong disapproval, passing events were regarded by the Hindus as devoid of real significance and so it is not surprising that no importance was attached to providing them with accurate dates. The Hindus were much more interested in devising elaborate cosmic cycles of vast and terrifying proportions. With their love of large numbers, they assigned to a single cycle 12,000 divine years, each of 360 solar years, totalling 4,320,000 years, and 1,000 of such cosmic cycles constituted one *kalpa*. This was equal to but one day in the life of Brahma.[3]

Hindu thought concerning the nature of time is well illustrated by the way in which causal relations were expressed in Sanskrit. To indicate the causal relation between two notions, a compound was formed in such a way as to suggest that it is natural to begin with the effect and trace it back to its cause. This attitude tends to eliminate time, since both effect and cause are regarded as co-present in the mind. Any sequence of causally related phenomena is thus always regarded as complete. This retrospective way of thinking tended to be a general feature of thought in countries such as India and China. It can be contrasted with the thought processes of Western science, in which the march of phenomena is considered to have a definite and unique temporal direction from cause to effect.

China

The Chinese were more interested in the practical measurement of time than were the inhabitants of India. Although the clepsydra was not invented in China, it was in use at an early stage of Chinese history. It probably came there from Babylonia, where the simplest form of the outflow type had already been in use before the time of the early Shang period (*c.*1500 BC). The Chinese also knew another archaic type of water-clock, a floating bowl with a hole in its base that was adjusted so that it took a specific time to sink.[4] But from the Han time onwards, that is after about 200 BC, the inflow type predominated. It was soon realized that more than one reservoir was required in order to avoid the slowing down of timekeeping that occurs with the falling pressure-head in a single vessel. Various other improvements were made, including the use of mercury, which does not freeze in cold climates in winter. Elaborate water-clocks were constructed between the second and eleventh centuries AD culminating in the remarkable instrument designed, and erected in the year 1088, by Su Sung (1020–1101), a Chinese mandarin.

Although this clock has not survived, its description by its inventor has. It was rediscovered in the mid-1950s by the leading authority on Chinese science and technology, Dr Joseph Needham of Cambridge.[5] The essential feature of this timekeeping device was a linkwork escapement quite different from the verge-and-foliot system invented in Europe in the late thirteenth century (ch. 7). Water poured continually from a constant-level tank into one scoop after another of a large water-wheel, but each scoop could not descend until it was full. As it went down it tripped two levers or weighbridges which by means of linkwork connections released a gate at the top of the wheel so as to let it move on by just one scoop. In effect, this machine dissected time by the weighing of successive equal quantities of fluid. An astronomical check on time-keeping was made by a sighting tube pointed to a selected star. Since the timekeeping was governed mainly by the flow of water rather than the escapement action, this device can be regarded as a link between the

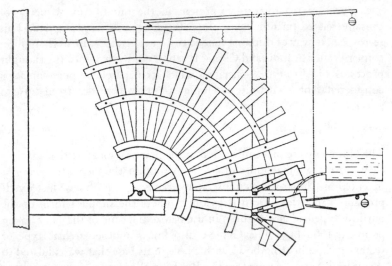

Fig. 1 A medieval Chinese water-clock. Although not a mechanical clock in the strict sense, Su Sung's water-clock involved a very early type of escapement. Each 24 seconds the weight of water that had poured at a steady rate into an empty cup became just sufficient to press it down and cause the wheel to rotate by one more spoke, thereby placing another empty cup under the spout. The wheel had 36 spokes and so made 100 rotations in 24 hours.

timekeeping properties of a steady flow of liquid and those of mechanically produced oscillations. This Chinese water-wheel clock was not only quite different from, but was a good deal more accurate than, the first European mechanical clocks and in this respect may not have been surpassed until after the introduction of the pendulum clock in the seventeenth century.[6] Despite its technological sophistication, the use of this elaborate clock was essentially astrological, the object apparently being to ensure that the positions of the celestial bodies would be known even if the Heavens were cloudy when any of the Emperor's wives or concubines produced offspring.

When the first European missionaries came to China in the sixteenth and seventeenth centuries no trace remained of the heavenly clockwork of five hundred years earlier. Indeed, the mechanical clocks which they presented to the Chinese rulers were received with surprised delight. Meanwhile, a different and much more widespread tradition of time-keeping survived involving the use of fire and incense. This tradition has been traced back to the sixth century AD. Because incense burns at an even rate and without flame, it is well suited for measuring the division of the religious day and for other purposes. Indeed, incense-clocks appear to have been as extensively used by the Chinese as sundials and clepsydrae. Both incense sticks and graduated candles were in common use for time measurement during the Sung dynasty (AD 960–1279), and both were later introduced into Japan. Silvio Bedini has commented that 'although the invention of the candle timepiece has been traditionally ascribed to Alfred the Great of England, it obviously had an earlier history in the Orient'.[7] Because it is inexpensive, burning incense-sticks to tell the time continued to be used down to the present century. In some of these clocks different pieces of incense gave off different aromas, thereby enabling those with sensitive noses to tell the approximate time.

The Chinese attitude to history was based on the belief that the rise and fall of dynasties was controlled by the mandate of Heaven. A man of modest birth might secure this mandate and become the Son of Heaven, or priest-king, but when his descendants displayed a lack of proper reverence for his example they lost the support of Heaven and their dynasty collapsed. Once the sign of Heaven was clear, there was then nothing disloyal in transferring allegiance to a new emperor, for he was blessed by the ancestral spirits of China. Thus in China the past had a definite social purpose, its use depending essentially on the concept of the mandate of Heaven to ensure continuity in a world of political change. J. H. Plumb has raised the interesting question 'Why did history develop in Europe, whereas in China it never extracted itself from the iron grip of the past in the service of the present?'[8] For, despite acquiring a mass of archival material stretching over a very long period of time, the Chinese never developed anything corresponding to the modern Western concept of history. In Plumb's opinion, '*What closed their minds to the historical problem was its absence.*' The European past, with its record of interaction between conflicting civilizations, religions, and cultures, lacked the unity and 'all-embracing certainty' of the Chinese and so presented historical problems of a kind the latter never encountered. Moreover, in China there was nothing corresponding to the secularization of history in Europe after the domination of the Church ended with the Renaissance and Reformation. China maintained bureaucratic control of both historical materials and their interpretation, whereas in Europe historical criticism had a much freer opportunity to develop.

As regards the philosophical concept of time, according to Needham, the Mohist school (followers of the philosopher Mo Ti in the fifth century BC) was inclined to temporal atomicity, although the hypothesis of material atomism never played any significant role in Chinese thought, which was wedded to the idea of the continuum.[9] More remarkably, the Mohists appear to have been near to formulating the concept of functional dependence in the relation of motion to time, an idea that was not fully developed until the scientific revolution of the seventeenth century in Europe. Generally speaking, however, different intervals of time tended to be thought of as separate discrete units. The universe was regarded as a vast organism undergoing a cyclic pattern of alternation, with now one and now another component taking the lead, the idea of succession being subordinate to that of interdependence. Just as space was decomposed into regions, time was split up into eras, seasons, and epochs. Consequently, as Needham has pointed out, in so far as Chinese natural philosophy 'was committed to thinking of time in separate compartments or boxes, perhaps it was more difficult for a Galileo to arise who should uniformise time into an abstract geo- metrical co-ordinate, a continuous dimension amenable to mathematical handling'.[10]

China was so remote from Europe that when silk first became available for the wealthier classes in Imperial Rome no one knew where it came from, but the fact that it reached Rome shows that these two civilizations were not totally devoid of contact with each other. In the Middle Ages a stream of inventions, such as gunpowder, paper, and the mariner's compass, passed from China to Europe.

The Maya

A civilization which paid particular attention to time but was destined, however, to remain totally isolated from both Europe and Asia until long after its decline—it reached its peak between about AD 600 and 900—was that of the Maya of Mesoamerica. The Maya were an agri- cultural people who had to contend with a capricious climate. They must have felt the need for a farmer's almanac and so started to keep a tally of days, which they recorded with special symbols. As regards their gods, benevolent ones were honoured, whereas those of an equivocal nature had to be appeased at the appropriate times. It was even more important to know when evil gods would be in charge so as to avoid trouble as far as possible by doing nothing at such times. In response to these needs a calendrical system of considerable complexity was devised. This was facilitated by the development of a remarkable notation for numbers, the Maya being among the most numerate civilizations there have ever been. They used the concept of place value and also had a symbol for zero.

Instead of a decimal system like ours or a sexagesimal system like the Babylonian, the Maya worked with a system of numeration based on the number twenty. There were twenty days in the month, each being regarded as divine and of distinctive omen. Thirteen months of twenty days gave a cycle of 260 days that formed the core of the Maya almanac. It has been suggested that the reason for this choice may have been that the longer of the two annual intervals between successive zenith-transits of the sun in the neighbourhood of Copan (one of the most important Maya cities) at latitude 15 degrees north is 260 days.[11] The 260 days of the sacred year each had a number from one to thirteen attached to it and there were also twenty different day-names, arranged so that the same combination of number and name only came round again 260 days later. The god of the day of the 260-day cycle on which a man was born was his patron saint or guardian deity. There is evidence that it became the practice for individuals to be named after the day on which they were born and that couples were not allowed to marry if their birthdays had the same numeral.[12]

Besides the 260-day cycle (or 'Sacred Year') the Maya had a solar year of 365 days, known to archaeologists as the 'Vague Year', composed of eighteen months of twenty days each and five intercalary days. The next largest cycle in the Maya 'Calendar Round' contained 18,980 days, corresponding to the period when the 260-day and 365-day cycles meshed together. The number 18,980 is the least common multiple of 260 and 365. This number of years is equal to 52 Vague Years and to 73 Sacred Years. The *katun*, comprising twenty years of 360 days, was the most important unit of time in the Maya view, because the events in one *katun* were expected to approximate to those in a previous *katun* that had ended on a day with the same number. A notable feature of the Maya calendar was the era known as the 'Long Count', a day-count which began from a conventional starting-point believed to be 10 August 3113 BC according to our calendar. This may perhaps have corresponded to the last creation of the world, for the Maya believed that the world had been created and destroyed several times. It is now thought that the 'Long Count' was used to date historical rather than astronomical events. Since about 1960 there has been a great increase in our knowledge of Maya history. Many inscriptions on monuments that were previously thought to be purely calendrical are now known to commemorate specific events of historical significance,[13] but the only three Maya books that have survived are all devoted to astronomy.

The most important of these is the so-called *Dresden Codex*, which includes a set of tables for the planet Venus that is remarkable for its accuracy.[14] This planet was identified with Kukulcan, the Maya equivalent of the sinister Mexican feathered serpent Quetzalcoatl. The prediction of its heliacal rising after inferior conjunction, that is, its first reappearance as 'the morning star' after a period of invisibility, was of vital concern to the Maya, who regarded it as a moment of particular dread. Every Maya cycle had its re-entry stage on a unique day in the 260-day Sacred Year, the day of Venus being 1 Ahau. The main problem was to determine after how many synodic revolutions, or reappearances of Venus as 'the morning star', this phenomenon would recur on 1 Ahau. If the synodic period of Venus were exactly 584 days, the number of revolutions required would be 65, corresponding to 146 of the 260-day cycles. This is because the least common multiple of 584 and 260 is 37,960, which is equal to the product of 65 and 584 and also to the product of 146 and 260. The Maya priests discovered, however, that 584 days is a slight overestimate of the average synodic period of Venus. (The correct result is 583.92 days, to the second place of decimals.) They coped remarkably well with this discrepancy. Despite there being no fractions in their arithmetic and observational difficulties due to frequent early morning mists and much cloud in the rainy season, they ultimately attained an accuracy in the determination of the synodic period of Venus that was of the order of one day in five thousand years, that is roughly one part in two million. This degree of accuracy was attained in European planetary astronomy only in modern times. It was nearly twice that of our present Gregorian solar calendar and almost forty times that of the Julian calendar of contemporary Europe. For such a remarkable achievement the close co-operation of many generations of patient observers must have been necessary.

Of all ancient peoples the Maya appear to have been the most obsessed with the idea of time. There is evidence that some earlier peoples in Central America, particularly the Olmec, were also concerned with time, but none to the degree that the Maya were.[15] Whereas in European antiquity the days of the week were regarded as being under the influence of the principal heavenly bodies—Saturn-day, Sun-day, Moon-day, and so on—for the Maya each day was itself divine. The Maya pictured the divisions of time as burdens carried by a hierarchy of divine bearers who personified the respective numbers by which the different periods of time—days, months, years, etc.—were distinguished. The burdens were carried on the back, the weight being supported by a tump-line across the forehead. There were momentary pauses at the end of each prescribed period when one god with his burden succeeded another. A graphic description of this theme has been given by one of the foremost experts on the Maya, J. E. S. Thompson:

One god raises his hand to the tump-line to slip it off his forehead, whereas others have slipped off their load, and hold them in their laps. The night god, who takes over when the day is done, is in the act of rising with his load. With his left hand he eases the weight on the tump-line; with his right hand on the ground he steadies himself as he starts to rise. The artist conveys in the strain reflected in the god's features the physical effort of rising from the ground with his heavy load. It is the typical scene of the Indian carrier resuming his journey familiar to anyone who has visited the Guatemalan highlands.[16]

Despite their constant preoccupation with temporal phenomena, the Maya never attained the idea of time as the journey of a single bearer with his load. Their conception of time was magical and polytheistic. The road along which the divine bearers marched in relays had neither beginning nor end. Events moved in a circle represented by the recurring spells of duty for each god in the succession of bearers. Days, months, years, and so on were all members of relay teams marching through eternity. Each god's burden came to signify the particular omen of the division of time in question. One year the burden might be drought, another a good harvest, and so on. By calculating which gods would be marching together on a given day, the priests could determine the combined influence of all marchers and thus forecast the fate of mankind. The hierarchy of cycles for each division of time led the Maya to devote more attention to the past than to the future. For although the particular details might vary, history was expected to repeat itself in each *katun*, and significant events would follow the pre-ordained general pattern. Thus, in the Maya world-view there was no sense of progress but only a blending of past, present, and future, which all tended to become one.

Classic Maya civilization appears to have collapsed some 600 years before the Spanish conquest of Central America, but even if it had survived it seems inevitable that the Maya obsession with time would have remained a historical curiosity with no influence on the modern world. For, despite the remarkable mathematical and astronomical expertise displayed by the Maya priests, their mental outlook was magical and not scientific. They were far more interested in time than were the Greeks, but the fact that every moment was regarded as the manifestation of supernatural forces meant that the concept of time which dominated Maya thought was purely astrological. A civilization that never invented the wheel was automatically precluded from inventing the mechanical clock,[17] but in fact neither the sundial nor the water-clock appear to have been developed by the Maya for measuring the passage of time. In short, they had the calendar but not the clock.

The author of a recent book on the history of clocks and their influence on the modern world has remarked that anyone looking at the techniques of time measurement throughout the world in, say, the eleventh century 'would have given odds that the Chinese would develop a mechanical clock well before the Europeans'.[18] On the contrary, Chinese horological techniques did not advance, and when the Jesuits brought their clocks to China in the sixteenth century the inventions of Su Sung and others had long been forgotten. Like the precise astronomical observations of the Maya, these technical achievements proved to be a dead end. Instead, it was in western Europe that the mechanical clock first appeared and with it a new type of civilization based on the measurement of time.

Part III
Time in the Modern World

7. The Advent of the Mechanical Clock

The invention of the verge escapement

In antiquity the only mechanical (or, more strictly speaking, quasi-mechanical) instruments for recording the passage of time were water-clocks. The fundamental difference between water-clocks and mechanical clocks, in the strict sense of the term, is that the former involve a continuous process, for example, the flow of water through an orifice, whereas the latter depend on a mechanical motion that continually repeats itself and so divides time uniformly into discrete segments. Many of the ancient water-clocks were instruments of considerable complexity, particularly as they were designed to indicate hours which varied throughout the year. Although there were no mechanical clocks in antiquity, mechanical models appear to have been constructed to reproduce the relative motions of the heavenly bodies. Writing in the first century BC, Cicero (*De republica*, I. xiv. 22) referred to one invented in Syracuse by Archimedes (287–212 BC). We know nothing about the gearing devices involved, but the associated mathematical calculations may have been contained in a lost treatise of Archimedes *On sphere-making*, that is on modelling the heavens. One remarkable Hellenistic geared mechanism, however, has survived from the first century BC. It was discovered in 1900 in the wreck of a Greek ship near the barren islet of Antikythera, off the south coast of Greece. In 1974, D. J. de Solla Price reported on the results of X-ray and gamma-ray radiography of the corroded remains of this bronze mechanism and concluded that it was a calendrical computing device.[1] It appears to have included means of determining the positions of the sun and moon in the zodiac, and it involved an assembly of wheels with fixed gear-ratios for the mechanization of the Metonic cycle, in which 19 solar years correspond to 235 lunar months (see Appendix 2). According to our present knowledge, this machine was the nearest the artificers of antiquity came to inventing a truly mechanical clock.

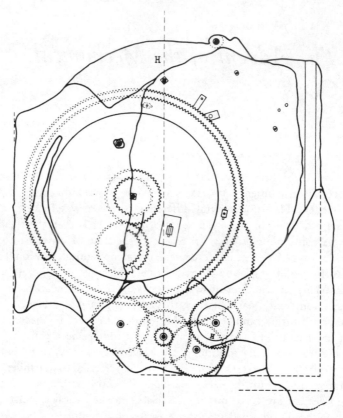

Fig. 2 Reconstruction of the Antikythera geared mechanism. This is a diagram of the differential gear assembly of the Antikythera mechanism as reconstructed by D. J. de Solla Price from the four surviving heavily corroded bronze fragments, which appear to contain the remains of 31 gear wheels. From the respective numbers of teeth in these remains, he claimed that the function of the instrument was calendrical, since it seemed to him that in the complete mechanism the numbers 19 and 235 of the Metonic cycle (Appendix 2) were involved. It is possible that this mechanism formed part of a collection of 'spoils of war' that was being conveyed by sea to Rome following the sack of Athens by Sulla's troops in 86 BC.

Until recently the Antikythera mechanism was thought to be the sole surviving example of mathematical gearing in the Hellenistic tradition. However, in 1983 four fragments of a geared instrument of early Byzantine origin, probably made during, or just before, the reign of Justinian I (527–65), were acquired by the Science Museum, London.[2] It has been possible to reconstruct the complete instrument, which was a brass portable sundial with a geared calendar that showed the approximate shape of the moon and its age in days and may also have shown its position and that of the sun in the zodiac. Two of the fragments involve gears of fifty-nine and nineteen teeth and of ten and seven teeth, respectively. These correspond to parts of a mechanical calendar described by the Persian scientist al-Biruni (973–1048) about the year 1000. A practical link has thus been revealed between the Hellenistic tradition of mathematical gearing and the medieval Islamic. The only surviving example of the latter is a calendrical mechanism yielding the shape and age of the moon in days and the positions of the sun and moon in the zodiac. It is attached to a Persian astrolabe of the early thirteenth century now in the Museum of the History of Science at Oxford. The earliest surviving gearing from the Latin West is attached to a French astrolabe of about 1300, now in the Science Museum, London.

Although no definite links have yet been discovered between the first mechanical clocks and earlier geared astronomical models and automata, the way in which a surviving late fourteenth-century clock such as that of Wells Cathedral displays the phases of the moon and figures which emerge at successive hours suggests that such clocks were the product of a continuing tradition from the distant past. There is also textual evidence to support this view. Nevertheless, the actual origin of the mechanical clock remains a mystery, although it probably occurred towards the end of the thirteenth century.

Early that century the market for water-clocks was such that a guild of clockmakers is known to have existed in Cologne who by 1220 occupied a special street, the Urlogengasse, or Clockmakers Street.[3] Since, in northern climes, water-clocks must have been a nuisance in winter when they froze, in the fourteenth century sand-clocks were invented. This invention was made following the introduction of a new and finer 'sand' made of powdered eggshell. Coarse sand cannot be used for this purpose, because it soon enlarges the hole through which it flows. Sand-clocks proved suitable only for measuring short periods. They were principally used on board ship to measure its speed by counting the number of knots paid out on a line tied to a log floating astern, while the sand-glass measured a given time that was usually half a minute. They were also used for timing the length of sailors' watches. Incidentally, it was not until the end of the fifteenth century that the sand-glass was depicted as the attribute of Father Time.[4]

The incentive to develop the mechanical clock may well have been fostered by the need for it in medieval monasteries, where punctuality was a virtue that was rigorously insisted on and late arrival at divine service or meals was punished. The need for punctuality was not due to any desire for 'saving time' but because the strict regulation of time was needed to help maintain the discipline of monastic life. In any case, it seems inevitable that the development of the mechanical clock should have been primarily due to the Church for, although the transmission of power by rope and pulley had long been known to craftsmen, the mathematics of gear-trains (particularly astronomical trains) was known only to the highly educated, and their education was provided only by the Church.

The English word 'clock' is etymologically related to the medieval Latin word *clocca* and the French word *cloche*, meaning a bell. Bells played a prominent part in medieval life, and mechanisms for ringing them, made of toothed wheels and oscillating levers, may have helped to prepare for the invention of mechanical clocks. Possible evidence for this view can be seen in the only surviving thirteenth-century picture of a Western water-clock, which seems to have been used about 1250 in Paris at the court of Louis IX. It was essentially a device for ringing the hours. The only visible wheel appears to have twenty-four teeth, which may signify that it rotated daily. The driving power was provided by a slowly descending weight hanging from a cord wound round the axle, this being the earliest instance known of a weight drive in a clock. It was followed about twenty years later, in 1271, by the *forecast* of a purely mechanical chronometer by Robertus Anglicus ('Robert the Englishman') in a commentary that he wrote on the *Treatise on the Sphere* of Sacrobosco. He envisaged this as a well-balanced wheel driven by a lead weight suspended from its axle so that it would make one revolution between sunrise and sunset. Nevertheless, he said of those clockmakers who were attempting to make such a timepiece that 'they cannot quite perfect their work'.[5]

Under the patronage of Alfonso X (*el Sabio*, 'the wise') of Castile a set of improved astronomical tables, known as the 'Alfonsine Tables', was compiled by Rabbi Isaac ben Sid of Toledo and published in the *Libros del saber de astronomica* in 1277. In Volume IV of this work, republished in Madrid in 1866, various inventions are described including a 'mercury clock'. This is a weight-driven clock equipped with a brake consisting of a drum divided into twelve compartments with small holes in the dividing walls. The lower six compartments are filled with mercury. As the driving weight causes the drum to rotate, the mercury is raised until it counter-balances the weight, which can then fall slowly as the mercury flows through the dividing walls. The uniform motion of the drum depends on the viscosity of the mercury. The motion can be regulated by varying the weight and/or the size of the drum. The essential feature of the mechanical clock missing from this interesting device was the 'escapement'. The invention of the mechanical clock probably occurred after 1277, since if it had occurred earlier it is almost certain that it would have been included in Volume IV of the *Libros del saber de astronomica*. It would seem that the date of the invention of the mechanical clock is probably some time between 1280 and 1300.

The crucial invention that made the mechanical clock possible was the 'verge-and-foliot' escapement. A horizontal bar, or 'foliot', was pivoted at its centre to a vertical rod, or 'verge', on which were two pallets. These engaged with a toothed wheel (driven by a weight suspended from a drum) which pushed the verge first one way and then the other,

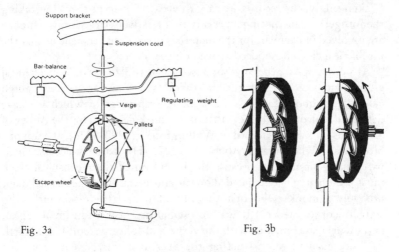

Fig. 3a   Fig. 3b

Fig. 3 The verge-and-foliot mechanical clock. The *foliot* (from the Latin word for 'leaf') was a horizontal bar (or balance) with a weight (or regulator) at each end (Fig. 3a). At its mid-point the bar was fixed to the *verge* (from the Latin word for 'twig'), a vertical rod on which were two pallets (or flanges). These engaged with a toothed wheel, which pushed the verge first one way and then the other (Fig. 3b), thereby causing the foliot to oscillate. The wheel itself advanced by one tooth for each double oscillation. The rate of oscillation could be adjusted by altering either the weights or their distance from the verge. This ingenious mechanism was robust, almost impervious to wear, and capable of ticking away ceaselessly so long as its moving parts were kept well-oiled. Its main disadvantage was that, unlike a pendulum, the balance controlling the oscillations had no natural period of its own.

causing the foliot to oscillate. The wheel advanced, or 'escaped', by the space of one tooth for each to-and-fro oscillation of the foliot. The foliot carried two weights (regulators) on each side, and the speed of the oscillation could be adjusted by altering either the weights or their distance from the verge. (In Italy the foliot was sometimes replaced by a balance wheel with a similar reciprocating action.) The system also involved a mechanism for counting the oscillations. No one knows who first made this ingenious invention, although as already indicated it was probably towards the end of the thirteenth century. According to C. F. C. Beeson, the earliest European record of a clock with a mechanical escapement is that of 1283 in the *Annals of Dunstable Priory* in Bedfordshire.[6] He also cites records from Exeter Cathedral (1284), old St Paul's, London (1286), Merton College, Oxford (1288?), Norwich Cathedral Priory (1290), Ely Abbey (1291), and Canterbury Cathedral (1292). As J. D. North, who has drawn attention to these, remarks, 'Taken singly, the records are easy to view with scepticism, but taking them together, and noting especially that relatively large sums of money are involved in payment for the materials used, they persuade us that the mechanical clock had indeed arrived on the scene.'[7]

Although it is generally supposed that the first truly mechanical escapement was of the verge-and-foliot type found in various church clocks throughout Europe, the earliest escapement of which we have definite detailed knowledge is that of the clock designed for the Abbey of St Albans *c.*1328 by Richard of Wallingford (*c.*1292–1336), the son of a blacksmith, who became Abbot in 1327. (His father's occupation is particularly significant, because the invention of the mechanical clock must presumably have depended on the co-operation of the learned man, probably a monk, who first thought of it and the blacksmith who actually constructed it.) It was an oscillating mechanism involving an extra wheel, as compared with the verge-and-foliot system. J. D. North has succeeded in reconstructing the St Albans escapement from the purely verbal description given in the surviving manuscript.[8] It was in some ways superior to the verge type. North has also found that a similar escapement was known more than a century and a half later to Leonardo da Vinci. Drawings of it are given in his *Codex Atlanticus* of about 1495, but Leonardo can no longer be regarded as its inventor. The St Albans clock had two similar escapements, one to control the going-train and one to ring the bell each hour on a twenty-four-hour system with the number of strokes equal to the hour. According to North, it is not inconceivable that such an oscillating striking device, triggered at suitably chosen intervals by a hydraulic clock, pointed the way to the first mechanical escapement proper.

The oldest surviving clock in England is the Salisbury Cathedral clock that was made not later than 1386. It has no dial or hands but strikes the hours. The verge-period for a half-swing is four seconds. The clock was restored to its original condition in full working order in 1956, after a lapse of seventy-two years.[9] Another more complete clock, believed to be by the same craftsman, that was in Wells Cathedral from at least 1392 is now in the Science Museum, London. (Both clocks were in due course converted to pendulum clocks.)

The accuracy of all early mechanical clocks was low, because the foliot and wheel had no natural periods of their own and also because of the effects of friction. It was, however, an age when civilization was becoming more vigorous and the number and skill of metal workers were increasing. A tremendous craze developed for the construction of elaborate astronomical clocks. As a leading historian of medieval technology has remarked, 'No European community felt able to hold up its head unless in its midst the planets wheeled in cycles and epicycles, whilst angels trumpeted and countermarched at the booming of the hours.'[10] Outstanding among these 'clocks' was the astrarium of Giovanni de' Dondi of Padua, designed between 1348 and 1364. This complex instrument with its finely cut teeth and intricate gearing was made of brass and was smaller than the clumsy early English clocks that were made of forged iron. It was only incidentally a timepiece. Primarily it was a mechanical representation of the universe, a kind of planetarium. It was much more elaborate than the first of the famous series of astronomical clocks in Strasbourg Cathedral that was installed about the same time, 1350. The original Strasbourg clock probably contained, besides moving figures, an annual-calendar dial, and possibly a lunar dial and an astrolabe, but the instrument designed by Giovanni de' Dondi incorporated a perpetual calendar for all religious feasts, both fixed and movable, and also indicated the celestial motions of the sun, moon, and planets, including even the motions of the nodes of the moon's orbit, which take over eighteen years to make a complete revolution around the ecliptic.

It is not surprising that this remarkably complete astronomical clock attracted the attention of princes. It was acquired in 1381 by Duke Gian Galeazzo Visconti, an intellectual who has been described as 'a sedate but crafty ruler with a great love of order and precision'.[11] He moved it to his palace in Pavia, where in 1420 it was recorded as being in the ducal library. It was very difficult to keep in working order, and when the

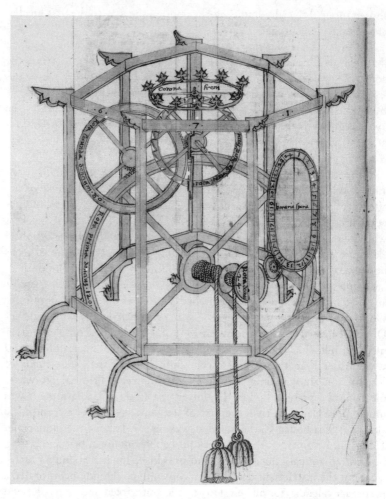

Fig. 4 A drawing of de' Dondi's astronomical clock. This drawing, dated 1461, of a part of de' Dondi's astronomical clock is from a manuscript in the Bodleian Library, Oxford. (MS. Laud Misc. 620, fol. 10ᵛ.) This complex weight-driven instrument, which was completed at Padua in 1364, was regulated by a horizontal balance-wheel shaped like a regal crown. Two pallets were fitted to the verge of this wheel. The upper one was made to move by one of the 24 teeth of the escape-wheel, which turned the verge and balance-wheel so that this pallet then escaped and at the same time the lower pallet became engaged. The latter then turned the verge and balance-wheel the opposite way, and so the process continued. The balance-wheel had a beat of 2 seconds. In recent years a number of models of this clock have been made; one is in the Museum of the Smithsonian Institution, Washington, DC.

emperor Charles V saw it in Pavia in 1529 it needed repair. Charles, who had a taste for mechanical devices, commissioned Gionallo Torriano of Cremona to repair it, but owing to corrosion he found that this was impossible and he agreed to make a similar instrument. When Charles retired to the monastery of San Yuste in 1555, with a large collection of clocks and watches, he took Torriano with him. After Charles V died in 1558, Torriano entered the service of his son Philip II and moved to Toledo, where he died in 1585. A few years ago manuscript evidence was discovered that Torriano's copy of de' Dondi's clock was still at his house in Toledo in the seventeenth century, and it is therefore unlikely that, as formerly thought, it perished when the convent of San Yuste with its art treasures was set on fire by the French in 1809. In recent years a number of working reconstructions of de' Dondi's clock have been made.

In the course of the fourteenth century mechanical clocks became progressively more numerous in Europe, most of those that were not installed in churches being public clocks. One such was a striking clock designed by Giovanni de' Dondi's father Jacopo, on whom the surname 'del Orologio' was conferred. It was erected in the entrance tower of the Carrara Palace at Padua in 1344, but was destroyed in the assault on that town by the Milanese in 1390. Although they were expensive, public clocks were generally regarded as being very useful. Whereas church bells announced the times of the various religious offices, the communal clock was a secular instrument that struck the hours, and by the end of the fourteenth century some were made that struck the quarters, although this did not mean that they were any more accurate. They were often unable to keep time to within fifteen minutes a day and were frequently out of order. This is not surprising since all geared wheels had to be cut by hand.

The social influence of the mechanical clock

An important consequence of the introduction of mechanical clocks was that in much of western Europe it led to the adoption of the uniform hour of sixty minutes. The earliest recorded clocks, such as the St Alban's clock and that erected in the Palace Chapel of the Visconti in Milan in 1335, struck up to twenty-four. Dante would appear to have seen a striking clock at least fifteen years before the Visconti clock of 1335 was installed; he may have seen the iron clock placed in the campanile of the church of Sant' Eustorgio in Milan in 1309—the first Italian public clock of which we have knowledge. In *Paradiso*, composed between 1317 and 1320, he made a famous reference (24. 13–15) to the striking-train of a clock:

> E come cerchi in tempra d'oriuoli
> si giran si che il primo, a chi pon mente,
> quieto pare, e l'ultimo che voli . . .

('And even as wheels in harmony of clockwork so turn that the first, to whoso noteth it, seemeth still, and the last to fly . . .')

Despite the inconvenience of counting large numbers the twenty-four-hour system persisted for centuries in Italy, but most other countries of western Europe soon adopted the system in which the hours were counted in two sets of twelve from midnight and from noon, respectively. The uniform hour of sixty minutes soon tended to replace the day as the fundamental unit of labour time in the textile industry. For example, in 1335 the governor of Artois authorized the inhabitants of Aire-sur-la-Lys to build a belfry whose bell would chime the working hours of textile employees.[12] The problem of the length of the working day was particularly important in the textile industry, where wages were a considerable part of production costs.

Despite the invention of the mechanical clock, for most people time remained uneven in quality. Nothing had done more to encourage this belief than the Church with its ecclesiastical calendar and regulations concerning what could or could not be done on specific days. Rules, or canons, were also established for the recital of prayers at definite times of the day. Known as the Canonical Hours, these followed the system of seasonal hours: Matins before dawn, Prime at sunrise, Tierce at 3, Sext at 6, Nones at 9, Vespers at 11 (the last four being reckoned from sunrise), and Compline after sunset. In due course Nones was set back three hours to midday, and this is the origin of the word 'noon'. Devout lay folk who wished to participate in this daily programme, however, needed their own prayerbooks. A 'Book of Hours' was the name given to such a prayerbook intended for private or family devotion, the term 'hours' indicating not an interval of sixty minutes, but less precise parts of the day that were set aside for religious and other duties. Originally, books of this type were commissioned only by kings and the highest nobility, but by the fifteenth century secular workshops had been set up, particularly in Paris and other cities in France and the Low Countries, so as to provide such books for a wider public. They form the largest single category of medieval manuscripts that have come down to us, and all later prayerbooks derive from them.[13] They usually begin with pictures of the occupations of the different months, followed by passages from the Gospels and the liturgical hours from Matins (and Lauds) to Vespers and Compline, and completed by miniatures of the life of the Virgin. The most celebrated of all these books is the *Très riches heures*, painted early in the fifteenth century by the Limbourg brothers for the Duc de Berry, third son of King John II of France.[14]

Popular superstitions concerning lucky and unlucky days, the existence of which has already been referred to in connection with the Romans, were reinforced in the Middle Ages by the recognition of black-letter days in the ecclesiastical calendar. For example, the day commemorating the massacre of the Holy Innocents, 28 December, was regarded as a day of particular ill-omen, especially in the fifteenth century. Moreover, throughout the year the particular day of the week on which Innocents' Day had fallen the previous year was also regarded as a black-letter day, and was also called Innocents' Day. Those who were influenced by this belief refrained from setting out on a journey or from starting a major task on that day of the week. An interesting example of this superstition concerns the coronation of Edward IV on 4 March 1461. In the preceding year 28 December fell on a Sunday and as the coronation took place on that day of the week it had to be repeated on another day![15] Even as late as the last quarter of the sixteenth century Queen Elizabeth's chief minister Lord Burghley warned his son to avoid undertaking new enterprises on three particularly ominous anniversaries in the ecclesiastical calendar: the first Monday in April (the murder of Abel), the first Monday in August (the destruction of Sodom and Gomorrah), and the last Monday in December (the birthday of Judas Iscariot).

A class of special days (and weeks) that has survived in both the Roman and Anglican Churches is that of 'Ember Days' and 'Ember Weeks'. Associated with the four seasons, the weeks concerned begin, respectively, on St Lucy's Day (13 December), the First Sunday in Lent, Whitsunday, and Holy Cross Day (14 September). Wednesdays, Fridays, and Saturdays in these weeks are the Ember Days. Traditionally, these days were set apart for special prayer and fasting. The following Sundays, for example Trinity Sunday, are the days specially fixed for the ordination of the clergy. This ancient custom was finally established as a law of the Church, *c*.1085, by Pope Gregory VII.

In England, belief in the uniformity of time was greatly influenced by the Puritans in their strong opposition to the practices of the Roman Church, in particular to the idea of special days in the ecclesiastical calendar. Instead, the Puritans advocated a regular routine of six days of work followed by a day of rest on the Sabbath, the famous Non-conformist ethic. During the course of the seventeenth century, despite the reaction against Puritanism which followed the restoration of the monarchy in 1660, this point of view became increasingly influential, so that by the end of the century it had come to be generally accepted. As Keith Thomas has pointed out, 'This change in working habits constituted an important step towards the social acceptance of the modern notion of time as even in quality, as opposed to the primitive sense of time's unevenness and irregularity.'[16]

In France, a step towards ending the dominance of the liturgical practices of the Church was taken as early as 1370 by King Charles V when he ordered all the bells in Paris to be regulated by the recently installed clock of the royal palace, designed by Heinrich von Wiek (Henri de Vic), and to be rung at hourly intervals. Although the practical difficulties of time measurement were such that until the middle of the seventeenth century most clocks had but one hand and the dial was divided only into hours and quarters, the abstract framework of uniformly divided time gradually became the new medium of daily existence.

This important development, which began in the towns, was fostered by the mercantile class and the rise of a money economy. As long as power was concentrated in the ownership of land, time was felt to be plentiful and was primarily associated with the unchanging cycle of the soil. With the increased circulation of money and the organization of commercial networks, however, the emphasis was on mobility. Time was no longer associated just with cataclysms and festivals but rather with everyday life. It was soon realized by many of the middle class that 'time is money' and consequently must be carefully regulated and used economically. As Lewis Mumford has pointed out, 'Time-keeping passed into time-saving and time-accounting and time-rationing. As this took place, Eternity gradually ceased to serve as the measure and focus of human actions.'[17]

A typical instance of late medieval anxiety about time occurs in a letter of 1399 written by the wife of the 'Merchant of Prato', Francesco di Marco Datini, to her ageing husband: 'In view of all you have to do, when you waste an hour, it seems to me a thousand. . . . For I deem naught so precious to you, both for body and soul, as time, and methinks you value it too little.'[18] Two years later we find Datini himself writing in the same vein to one of his partners in Spain, Cristofano di Bartolo, whom he wished to persuade to come home. A similar note was struck by Chaucer in the Host's introduction to the 'Man of Law's Tale' in *The Canterbury Tales*, written about 1400:

> And therefor by the shadwe he took his wit
> That Phebus, which that shoon so clere and brighte,
> Degrees was fyve and fourty clombe on highte;
> And for that day, as in that latitude,
> It was ten of the clokke, he gan conclude,
> And sodeynly he plighte his hors aboute.
> 'Lordinges,' quod he, 'I warne yow, al this route,
> The fourthe party of this day is goon;
> Now, for the love of god and of Seint John,
> Leseth no tyme, as ferforth as ye may;
> Lordinges, the tyme wasteth night and day,
> And steleth from us, what prively slepinge,
> And what thurgh necligence in our wakynge,
> As dooth the streem, that turneth never agayn,
> Descending fro the montaigne into playn.'

Before long there were many activities for which time came to be increasingly regarded as valuable. In the early and high Middle Ages it had been possible to spend many tens and even hundreds of years on erecting a single building, be it a cathedral, a castle, or a town hall. This was possible because human life was regarded as primarily the life of the community in which one generation quietly succeeded another, so that there was no pressing need for rapid construction. All this was destined to change in the late Middle Ages and Renaissance period. Even in painting the time factor made itself felt. For although we frequently find in paintings of this period that a number of consecutive scenes are represented simultaneously in one picture, in other ways temporal considerations came to exert a decisive influence—in particular, causing painting *a secco* to replace *al fresco*, or true fresco, since the very long apprenticeship that pupils had to serve before they became proficient in fresco painting could not be maintained, and a successful painter had to work fast in order to handle all the commissions that he received. Even as great an artist as Michelangelo (1475–1564) was unable to turn the tide. Originally it had been planned that the Last Judgement in the Sistine Chapel should be painted *a secco* in oil, but he insisted on carrying out the work *al fresco* since he considered oil painting to be only 'fit for women and slovenly people'! His point of view conflicted with the spirit of the age and, despite his example, the glorious art of true fresco died out, its practice being incompatible with the new social attitude to time.

This attitude was also responsible for a new horological invention that was ultimately to be of far-reaching social significance. The first mechanical clocks were large and unwieldy, and there was soon a desire for smaller and more portable mechanisms. To meet this demand springs began to be used in the fifteenth century in place of weights as the source of motive power in clocks. This development was important because it made possible the invention of the domestic clock and also the watch. One of the earliest references to a watch is the gold pomander 'wherein is a clocke' that was presented in 1540 by Henry VIII to Catherine Howard, his fifth wife.[19] The public clock, whether installed in a church or in a town square, was only an intermittent reminder of the passage of time, but a domestic clock or a watch was a continually visible indicator. As D. S. Landes has pointed out, whereas the public clock could be used to open and close markets, to signal the start and end of work and to move people around, it signalled only moments rather than the continual passage of time. A chamber clock or watch, on the other hand, was an ever-visible reminder of 'time used, time spent, time wasted, time lost'. As such it was prod and key to personal achievement and productivity.[20]

Nevertheless, centuries were to elapse before this invention became widespread. Indeed, for a long while the possession of a domestic clock or a watch tended to be restricted to the wealthy and was looked upon more as a sign of affluence than as a social necessity. As late as the middle of the seventeenth century we find, for example, that even at the age of 30 Samuel Pepys (1633–1703), already an important government official, did not possess a watch. Instead, he lived by the church bells of London and occasionally a sundial, as did almost everyone else there. Consequently, very few specific appointments were made. Pepys moved around from public places to coffee houses and taverns hoping to do business. He often went to discuss matters with the Lord High Admiral, James, Duke of York, only to find that the Duke had gone hunting. Pepys never expresses surprise or resentment. Time had a different significance for him and most of his contemporaries than it has for us.

Since watches were for long the toys of the rich, it is not surprising that often when ordinary folk encountered one they were extremely puzzled and were even inclined to look upon it as something evil and dangerous. An amusing instance of this is related by John Aubrey concerning an Oxford don, Thomas Allen (1542–1632), who owned many mathematical and other scientific instruments. When staying one Long Vacation at a friend's place, 'at Hom Lacey in Herefordshire', he happened to leave his watch on the window-sill of his chamber. According to Aubrey,

The maydes came in to make the bed, and hearing a thing in a case cry *Tick, Tick, Tick*, presently concluded that it was his Devill, and took it by the string with the tongues, and threw it out of the windowe into the mote (to drown the Devill). It so happened that the string hung on a sprig of an elder that grew out of the mote, and this confirmed them that 'twas the Devill. [Consequently], the good old gentleman got his watch back.[21]

When most people possessed no clocks or watches but lived more in the countryside than they do today, they took far more note than we do of the various timings associated with plants and animals. Indeed, some plants were even named thereby, for example, the 'day's eye' (daisy), so-called in allusion to its revealing its yellow disc in the morning and concealing it again in the evening. Most notable of natural timings was cock-crow, and a famous tribute to the cock Chauntecleer's horological skill was paid by Chaucer in the 'Nun's Priest's Tale':

> Wel sikerer was his crowying in his logge,
> Than is a clokke, or an abbey orlogge.
> By nature knew he ech ascenscioun
> Of equinoxial in thilke toun;
> For whan degrees fiftene were ascended,
> Thanne crew he, that it mighte nat ben amended.

Although watches were extremely rare before the late seventeenth century, the influence of mechanical timekeeping had already made itself felt in a variety of ways, besides those already mentioned. By the sixteenth century mining operations had become closely regulated by the clock according to Agricola (Georg Bauer), who in his *De re metallica*, of 1555, noted the precise times of shifts. Many professional people, such as judges and teachers, commenced their duties at stated hours, and by the late Middle Ages even the often unruly undergraduates at universities such as Oxford were subjected to the discipline of fixed timetables. Lectures often began as early as 5 or 6 a.m. in the summer (7 a.m. in the winter), and sometimes the first lecture went on for three hours, no provision being made for any food before 10 a.m.[22] It is interesting to

trace the way in which the times of meals have changed over the centuries, particularly because in everyday life it is not just the clock which
tells us which part of the day we are in but the meals that we eat. Thus, in
common parlance, 'afternoon' now begins an hour or more after 'noon'
according to the clock. Also dinner has tended to get later and later in the
day. The fascinating details of this historical trend have been described by
Arnold Palmer.[23] Although, the rule of the clock affected most people in
the sixteenth century far less than it does us today, it was already sufficient to provoke Brother Jean in the *Gargantua* (1535) of Rabelais to
complain that 'the hours are made for man and not man for the hours!'[24]

8. Time and History in the Renaissance and the Scientific Revolution

Reform of the calendar

Throughout history the ultimate standard of time has been derived from astronomical observations. In due course this led to the hour, minute, and second being defined as fractions of one rotation of the earth on its axis. Since it was found convenient in everyday life to determine this rotation by the orientation of the earth relative to the sun, the 'mean solar day' was defined as the period of rotation of the Earth on its axis relative to the Sun corrected for all known irregularities. Because the earth's orbit is only approximately circular, the relative speed of the sun is not quite uniform. Also the sun's apparent motion in the sky is not along the celestial equator (the projection of the earth's Equator on to the sky), and consequently the component of the sun's motion parallel to the Equator varies. As a result, for the purposes of ordinary timekeeping a 'mean sun' is defined as moving at a constant rate which is the average of that of the actual sun. The difference between mean solar time and apparent solar time (as given by a sundial) is called the 'equation of time'. The 'equation of time' vanishes four times a year, on or about 15 April, 15 June, 31 August, and 24 December. The maximum amount by which apparent (or sundial) noon precedes mean noon is about 16.5 minutes on or about 3 November, and the maximum amount by which mean noon precedes apparent noon is about 14.5 minutes on or about 12 February. The 'mean solar second' is defined as the appropriate fraction (1/86,400) of the mean solar day.

Although in everyday life we find it convenient to determine time by the position of the earth relative to the sun, in practice it is more accurate to determine the times when stars cross the meridian, which is the projection on to the sky of the circle of longitude through the place on the earth's surface where the observations are being made. The interval between successive transits of the same star, or group of stars, across the meridian is called the 'sidereal day'. Since there is one more sidereal day in the year than there are solar days, the solar day is about four minutes longer than the sidereal day, which can be converted into the solar day by a numerical formula given to ten decimal places by observations extending over two hundred years.

The unit of time on which the seasons and the calendar depend is called the 'tropical year'. It is the time between two successive passages of the sun through the point at which its annual path against the general background of the stars (the 'ecliptic') crosses the celestial equator at the spring equinox. It is not the same as the time between successive passages of the sun through the same fixed point on the sky, because the equinox has a retrograde motion of just over 50 arc-seconds a year. This 'precession of the equinoxes', as it is called, is due to the gravitational pull of the sun and moon on the earth's equatorial bulge, thereby causing the earth's axis to precess, like the axis of a spinning top, with a period of about 25,800 years. The discovery of the precession of the equinoxes was made in antiquity by the Greek astronomer Hipparchus, and correct knowledge of it is required for the precise determination of the calendar. In antiquity the length of the tropical year was determined with the aid of a gnomon at noon at successive summer solstices (at the same place) when its shadow is shortest; whereas the length of the sidereal year was obtained from successive heliacal risings of the same bright star. According to modern measurements the tropical year is equal to about 365.2422 mean solar days, and the sidereal year to about 365.2564 of these days.

The estimate of the tropical year that was the basis of the Julian calendar, 365.25 days, was just over eleven minutes too long, equivalent to an extra day every 128 years. Consequently, by 1582 the spring equinox, which in Julius Caesar's day fell on 25 March, had retrograded to 11 March. Moreover, Easter, which by the decision of the Council of Nicaea in the year 325 should be celebrated on the first Sunday *after* the spring full moon (i.e. the full moon occurring on, or immediately after, 21 March) had got steadily further away from the full moon. To bring the equinox to 21 March, Pope Gregory XIII, acting on the advice of a special commission that included the distinguished Jesuit Papal astronomer Christopher Clavius, directed that the day after 4 October 1582 be designated 15 October (October was chosen because it was the month with the fewest saints days and other special ecclesiastical days), and that the leap year intercalary day be omitted in all centenary years except those that are multiples of 400. Thus 1600 was a leap year and

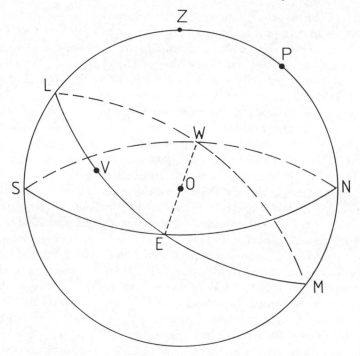

Fig. 5 The celestial sphere. Since they are so remote, the so-called 'fixed stars' (as distinct from planets, satellites, comets, and so on) can be depicted as fixed points on a sphere with a radius that is very large compared with the radius of the earth's annual orbit around the sun. The lines from all observers on earth to any given star will therefore cut this sphere, known as the 'celestial sphere', at the same point. Consequently, any observer O can be regarded as being at the centre of the celestial sphere. Because of the earth's diurnal rotation, the celestial sphere appears to rotate daily about the line joining the earth's north and south poles. In accordance with standard mathematical terminology, any circle on the celestial sphere which has its centre at O is called a 'great circle'. (Lines of longitude on the earth's surface are also great circles, whereas lines of latitude, except the equator, are not, being only so-called 'small circles'.) In the above figure the great circle ENWS represents the horizon of O, with its pole at the zenith-point Z. Similarly, the great circle ELWM represents the celestial equator (projection of the terrestrial equator on to the celestial globe). Its north pole is at P, the so-called 'Pole Star' being close to it. V represents the vernal equinox (first point of Aries). The ecliptic (projection on to the celestial sphere of the sun's apparent annual path against the general background of the stars, due to the earth orbiting the sun) is the great circle through V that lies in the plane cutting the plane of the celestial equator at an angle of approximately 23½ degrees. The earth's axis of diurnal rotation is in the direction OP. The celestial sphere rotates about the line joining O to the poles of the ecliptic in a top-like motion. This 'precession of the equinoxes' has a period of nearly 26,000 years.

2000 will be too, but the intervening centenary years were not. Moreover, it was decreed that the year should begin on 1 January. The new
calendar, had been suggested by a medical lecturer in the University of
Perugia, Luigi Giglio (latinized as Aloisius Lilius). Lilius died in 1576,
but his brother Antonio presented his scheme to the Pope. Unfortunately, Lilius's manuscript was never printed and is now lost. Clavius
regarded Lilius as a man entitled to immortality because 'he was the principal author of such an excellent correction.' For a more detailed account
of these questions and of the dating of Easter, see the Appendix.

At first only Catholic countries adopted the Gregorian calendar, since
in Protestant lands, despite some influential support for it, the feeling
became widespread that the Pope 'with the mind of a serpent and the
cunning of a wolf' was stealthily seeking by means of the calendar to
dominate Christendom once again.[1] Although this point of view now
seems ludicrous, it was not thought so at the time. For Gregory XIII was
not only a powerful promoter of the Counter-Reformation but had fully
supported Philip II in his ruthless campaign against the Protestants in the
Spanish Netherlands and had celebrated the St Bartholomew's Day
massacre of the French Huguenots in 1572 by having a medal struck to
commemorate it. Nevertheless, there were Protestant astronomers,
notably Tycho Brahe and Kepler, who approved of the Gregorian
reform, although others felt that Clavius had not applied sufficient scientific rigour in his investigations concerning it. In 1613, at the Diet of
Regensburg, Kepler (who, in supporting Clavius, made the point that
'Easter is a feast and not a planet. You do not determine it to hours,
minutes and seconds') argued that the Gregorian calendar did not
involve the acceptance of a papal bull but only the results of calculations
by astronomers and mathematicians.[2] Nevertheless, the Protestant states
maintained their opposition until 1700, when most of them decided to
adopt the Gregorian calendar. In England and Ireland, however, anti-
Catholic feeling, which was as much political as religious, successfully
prevented its introduction for another fifty years, until the inconvenience
of using a different date from that employed in the greater part of Europe
could no longer be tolerated. Already in 1583, however, Queen
Elizabeth's favourite mathematician, astrologer, and secret agent, John
Dee, using data from Copernicus, had produced an eleven-day correction
which he claimed was more accurate than the Gregorian ten-day correction due to Clavius. An English mathematical committee, composed of
the astronomer Thomas Digges, Sir Henry Savile, and a Mr Chambers,
agreed with Dee but recommended, to his disgust, that it would be more

convenient in practice to adopt the same calendar as the Continent. Queen Elizabeth's ministers Burghley and Walsingham approved of Dee's plan, but nothing came of it because of the violent opposition of the bishops, who argued that the new calendar showed the influence of Papism. When at last, in September 1752, the change was made Dee's correction was adopted, 3 September becoming 14 September.

In England 25 December was taken as the beginning of the year during the Middle Ages until the latter part of the twelfth century, when 25 March was chosen instead. The Church decided to begin its year on that day (Lady Day) because it was the day of the Annunciation, being exactly nine months ahead of Christmas Day. In England the year beginning on 25 March was called the 'Year of Grace'. Although January appeared as the first month of the year in calendars and almanacs, all official documents followed the dating of the 'Year of Grace' until 1751. In that year the official year began on 25 March and ended on 31 December. From then onwards the official year began on 1 January. These changes were authorized by an Act of Parliament of 1750. Only a minimal change, however, was made in the tax year, which still ends on 5 April. That date in the new style calendar corresponded to 25 March in the old style calendar. In Scotland the year has begun on 1 January since 1600.

Confusion can easily arise when we try to compare dates between 1582 and 1752 according to the Julian calendar that prevailed in England with the corresponding dates in the Gregorian calendar used in some of the principal European countries. For example, it has sometimes been asserted that Cervantes died on the same day as Shakespeare. Unfortunately, this remarkable coincidence did not occur; Cervantes died in Madrid on Saturday, 23 April 1616, according to the Gregorian calendar already in use there, whereas Shakespeare died at Stratford-upon-Avon on Tuesday, 23 April 1616, according to the Julian calendar still current in this country, the corresponding Gregorian date being Tuesday, 3 May 1616, and so Shakespeare in fact outlived Cervantes by ten days.

The greatest opposition to the new calendar arose in the eastern Churches, and was forcibly expressed by the Patriarchs of Constantinople, Alexandria, and Armenia. Not until 1923 did the Orthodox Church in Greece, Romania, and Russia adopt it. The monks on Mount Athos (in north-eastern Greece) have still not accepted it. Nearly all the monasteries there adhere to the Julian calendar, which is now thirteen days behind the Gregorian. Moreover, at one monastery they still reckon the time of day according to the original Georgian style, with sunrise always occurring at twelve o'clock. Everywhere else on the 'Holy

Mountain' they follow the old Turkish system, with sunset at that time. This system is said to go back to the Byzantines and at least has the advantage that the traveller knows from his watch how many hours of daylight there are left.[3] The island of Foula, twenty miles west of Shetland, still retains the Julian calendar for its festivals, such as Christmas and Hogmanay.

Although civil time is based on natural phenomena, we have seen that not only religious but purely political considerations can influence the construction of a calendar, as in the case of ancient Rome. A much more recent instance of this occurred when, after deposing Louis XVI, the National Convention, or French Parliament, decided to introduce a completely new calendar. It was decreed that Year I should begin on what would otherwise have been 22 September 1792, the day the Republic was proclaimed. New names, such as *Germinal, Prairial*, and *Thermidor*, were devised by the dramatist Fabre d'Eglantine for the twelve new months of thirty days each, divided into three 'weeks', each of ten days. At the end of the year there were five days of festival called *Sansculottides*, or 'Trouser-days'. (The *culotte*, or breeches, was regarded as an aristocratic garment, and the common people wore trousers. *Sansculotte* was originally a derogatory term applied by the upper classes to their lower-class opponents.) The sixth extra day in leap year was to be 'The Trouser-day' when, according to Fabre, Frenchmen 'will come from all parts of the Republic to celebrate liberty and equality, to cement by their embraces the national fraternity, and to swear, in the name of all, on the altar of the country, to live and die as free and brave Trousermen.'[4] Fabre also devised names for each day of the year, many referring to fauna, flora, minerals, and agricultural implements. The new calendar, which was described by the American statesman John Quincy Adams as an 'incongruous composition of profound learning and superficial frivolity, of irreligion and morality, of delicate imagination and coarse vulgarity',[5] had a short life. It was officially discontinued by Napoleon, and on 1 January 1806 the French reverted to the Gregorian calendar, which despite its imperfections is still the most widely used calendar in the world.

The pendulum clock and the clocklike universe

Although medieval scholars were not, as a rule, concerned with machines, they became more and more interested in mechanical clocks, particularly because of their connection with astronomy. It was generally believed that a correct knowledge of the heavenly bodies and their motions was necessary for the success of most earthly activities. The theory of astral influences was accepted by most Christian thinkers until the seventeenth century. That is why medical students were required to study astronomy and astrology, so that a horoscope could be cast of the hour when the patient fell ill and the propitious hour for the appropriate treatment, such as surgery, be determined. A present-day relic of the influence of astrology on medicine is our use of the Italian word 'influenza' for the viral infection which was thought in former times to be due to a malevolent flow coming down to the sufferer from an evil star. Another etymological relic is the word 'disaster': originally this referred to the unfavourable aspect of a star (Latin *astrum*). Carlo Cipolla has drawn attention to the assertion by a writer in 1473 that the public clock in Mantua served the purpose of showing 'the proper time for phlebotomy, for surgery, for making dresses, for tilling the soil, for undertaking journeys and for other things very useful in this world.'[6] In particular, people believed that a star 'born' when it first appeared on the horizon influenced the life of a child coming into this world at that moment, and that a star just setting at the moment of a child's birth had implications for the circumstances of his or her death.

The invention of clockwork and its application to mechanical models of the universe, such as de' Dondi's, made a powerful impact on many minds. It is, therefore, not surprising that the clock metaphor came to be used in a variety of contexts. For example, Jean Froissart, in his poem 'Li orloge amoureus' (*c*.1380) presented an elaborate allegory in which various aspects of chivalrous love were compared with the different parts of a mechanical clock, the verge-and-foliot escapement being associated with the virtue of moderation, since self-control was the highest in the canon of virtues of the medieval knight.[7] No doubt Froissart had an actual clock in mind when he wrote this poem, and if so it may well have been Henri de Vic's clock at the royal palace in Paris. Sadly, that famous clock later became an object of derision, as is evident from the scurrilous rhyme:

> C'est l'horloge du Palais;
> Elle va comme ça lui plaît!

A particularly interesting indication of the way in which the invention of clockwork began to influence philosophical thought occurs in a treatise by Froissart's contemporary Nicole Oresme (1323–82) on the question of whether the motions of the heavenly bodies are commensurable or incommensurable. Part of the treatise is in the form of an allegorical debate between Arithmetic who favours commensurability and Geometry the opposite. Arithmetic argues that incommensurability and irrational proportion would detract from the harmony of the universe. 'For if anyone should make a mechanical clock would he not move all the wheels as harmoniously as possible?'[8] This is an early example of the mechanical simulation of the universe by clockwork suggesting, at least implicitly, the reciprocal idea that the universe itself is a clocklike machine.

This idea came to the fore in the scientific revolution of the seventeenth century. Early that century Kepler specifically rejected the old quasi-animistic magical conception of the universe and asserted that it was similar to a clock. Among others who drew the same analogy was Robert Boyle (1627–91). In a passage in which he maintained that the existence of God is not revealed so much by miracles as by the exquisite structure and symmetry of the world—that is by regularity rather than irregularity—he argued that the universe is not a puppet whose strings have to be pulled now and again but

it is like a rare clock, such as may be that at Strasbourg, where all things are so skilfully contrived, that the engine being once set a-moving, all things proceed according to the artificer's first design, and the motions . . . do not require the particular interposing of the artificer, or any intelligent agent employed by him, but perform their functions upon particular occasions, by virtue of the general and primitive contrivance of the whole engine.[9]

Boyle's words clearly imply a conception of nature from which all traces of the animistic world-view, such as was still evident at the beginning of the seventeenth century in Gilbert's book on the magnet, have been banished. In the development of the mechanistic conception of nature in the course of that century the mechanical clock played a central role. It was surely no coincidence that the greatest practitioner of the mechanical philosophy in its formative period, the Dutch scientist Christiaan Huygens (1629–95), who in the first chapter of his *Traité de la lumière* declared that in true philosophy all natural phenomena are explained 'par des raisons de mechaniques', was also responsible for converting the mechanical clock into a precision instrument.

This development was based on the discovery of a natural periodic process that could be conveniently adapted for the purposes of accurate timekeeping. As the result of much mathematical thinking on experiments with oscillating pendulums, Galileo (1564–1642) came to the conclusion that each simple pendulum has its own period of vibration depending on its length. (Historians of science now ascribe priority in this important discovery to the French scientist Marin Mersenne (1588–1648).) In his old age Galileo contemplated applying the pendulum to clockwork which could record mechanically the number of swings.

The first pendulum clock was based on the theoretical researches of Christiaan Huygens who, because of his astronomical observations, felt the need for a more exact timekeeper than had been previously available. In June 1657 the government of the United Netherlands granted to Salomon Coster of The Hague the exclusive rights for twenty-one years to make and sell clocks in that country based on Huygens's invention. Huygens discovered two years later that theoretically perfect isochronism (uniformity of oscillation) could be achieved by compelling the bob to describe a cycloidal arc. (A cycloid is the curve described by a point-like spot on a circular wheel that rolls without slipping along a straight line.) Great as was Huygens's achievement from the point of view of theory, particularly as set forth in his famous treatise *Horologium oscillatorium*, published in Paris in 1673, the practical solution of the problem of more accurate timekeeping came only with the invention of a new type of escapement.

Huygens's clock incorporated the verge type, but about 1670 a much improved type, the anchor type, was invented that interfered less with the pendulum's free motion. Although it is not clear who was responsible for this invention, John Smith in his *Horological Disquisitions* of 1694 attributed it to the London clockmaker William Clements. In this form of escapement, as a tooth of the scape-wheel escapes from the pallet at one end of the anchor, so a tooth on the other side engages with the pallet at the other end of the anchor. For satisfactory functioning, however, clocks incorporating the pendulum and the anchor type of escapement had to be placed on a level surface, and consequently in portable domestic clocks verge escapements were retained.

For those who were not astronomers the sundial remained the arbiter of local time against which clocks and watches were checked, although few common sundials in the seventeenth century were capable of showing time more accurately than to within half a minute at best. In the first comprehensive scientific treatise on the art of horology, William Derham's *Artificial Clockmaker* (first edition 1696), attention was drawn to the need to correct sundial readings for the effect of atmospheric refraction when the sun is low in the sky.

Although the earliest watches were driven by a spring, there was no

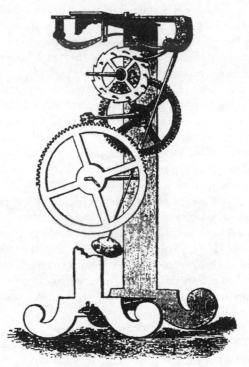

Fig. 6 A drawing of Galileo's pendulum clock. In 1637 Galileo devised a train of wheels actuated by a pendulum for counting oscillations, but the pendulum had to be controlled by hand. In 1641, the year before he died, Galileo considered how the pendulum itself could be used as a clock. In 1649 his son, Vincenzio, tried to construct a clock based on his father's design, but he died before completing it. (An inventory of his effects included an unfinished pendulum clock.) In 1659 a drawing by Galileo's friend and biographer Viviani of a clock based on Galileo's ideas was sent by one of his former disciples, Prince Leopold dei Medici, brother of the Grand Duke of Tuscany, to the French astronomer Ismael Boulliau. He passed it on to his friend Christiaan Huygens, who received it in January 1660. This drawing is reproduced above. Galileo's pendulum clock involved a new type of escapement which was superior to the traditional verge type retained by Huygens. Each swing of the pendulum pushes the top wheel from one projecting pin to the next.

controlling spring on the balance-wheel. Neither the foliot for clocks nor the balance-wheel for watches had a truly regular motion of their own and consequently no precise timekeeping property. Both the rate of swing of a pendulum under gravity and the motion of a balance-wheel under the control of a spring are, however, periodic. Just as the invention

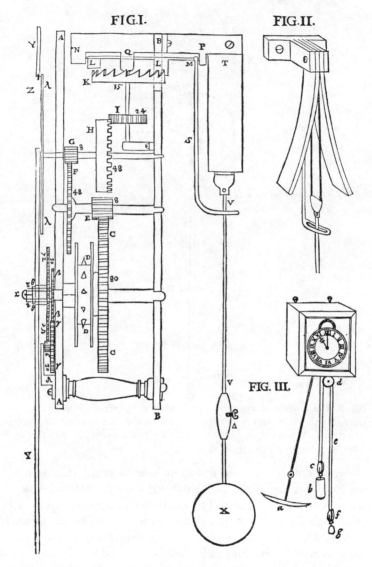

Fig. 7 Huygens's pendulum clock of 1673. These diagrams relating to Huygens's pendulum clock are on page 4 of his *Horologium Oscillatorium de Motu Pendulorum at Horologia Aptato*, published in Paris in 1673 'Cum Privilegio Regis' and dedicated to Louis XIV. Fig. I illustrates the works of the clock complete with verge escapement; Fig. II the cycloidal cheeks controlling the oscillations of the pendulum; and Fig. III the external appearance of the clock. In this clock, unlike Huygens's earlier clock, the pendulum was hung between cycloidal cheeks, so that its time of oscillation was independent of the size of the arc of swing—an important property for accurate timekeeping.

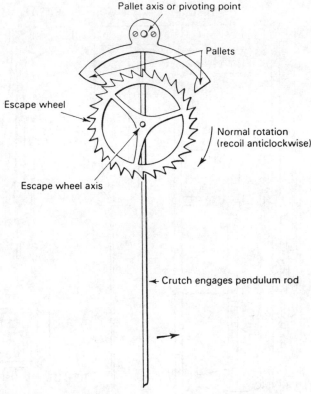

Pallet axis or pivoting point

Pallets

Escape wheel

Normal rotation
(recoil anticlockwise)

Escape wheel axis

Crutch engages pendulum rod

Fig. 8 The anchor escapement. The anchor escapement consists of a wheel with pointed teeth and an anchor carrying, at places equidistant from its axis, two pallets which catch the teeth of the wheel in succession as each escapes from the action of the other.

of the pendulum improved timekeeping by clocks, so the invention of the balance-spring about 1675 produced a similar improvement in the accuracy of watches. Robert Hooke (1635–1702) and Huygens each laid claim to the invention of the balance-spring. To Hooke can definitely be attributed the law of springs *ut tensio sic vis* ('the extension is proportional to the tension'), which he published in 1678 and which is named after him. Meanwhile, Huygens had actually made a spiral balance-spring, the idea of which Hooke claimed had first occurred to him but had been communicated to Huygens by Henry Oldenburg, the Secretary of the Royal Society, whom Hooke denounced as a 'trafficker in intelligence'! We can only conclude that, whereas Huygens definitely produced a watch with a balance-spring, there is abundant evidence that Hooke was the kind of ingenious inventor who often fails to follow up his insights sufficiently far to justify his claims. The question of Hooke's contributions to horology has been carefully examined by the historian of science Rupert Hall.[10]

There is no doubt that the achievement of greater precision in mechanical timekeeping in the second half of the seventeenth century was a momentous advance, for it ultimately led to recognition of the importance of precise measurement generally in science and technology. Moreover, the invention of an accurate mechanical clock had a tremendous influence on the concept of time itself. For, unlike the clocks that preceded it, which tended to be irregular in their operation, the improved mechanical clock when properly regulated could tick away uniformly and continually for years on end, and so must have greatly strengthened belief in the homogeneity and continuity of time. The mechanical clock was therefore not only the prototype instrument for the mechanical conception of the universe but for the modern idea of time. An even more far-reaching influence has been claimed for it by Lewis Mumford, who has pointed out that 'It dissociated time from human events and helped create belief in an independent world of mathematically measurable sequences: the special world of science.'[11]

We have seen that in referring to the famous Strasbourg clock Boyle said that 'the engine being once set a-moving, all things proceed according to the artificer's first design'. In the case of a clock, 'design' refers to the action of its mechanism and has no teleological significance. The mechanical conception of the universe was in this respect clocklike and in marked contrast to Aristotle's conception of the universe, which had greatly influenced medieval natural philosophers. That was based on the importance Aristotle attached to the fully developed forms to which, in his view, all things inanimate as well as animate aspire. Consequently, for him the essences or special qualities of things, rather than temporal sequences, were the primary objects of scientific investigation. This way of thinking came under fire in the seventeenth century, because it was increasingly felt that it failed to explain anything. Instead of postulating *ad hoc* qualities, scientists who rejected the views of Aristotle and his medieval followers invoked hypothetical mechanical systems to elucidate natural phenomena. In so far as such a system operates from given initial conditions it has some similarity to a clock. If a clock is to indicate the correct time, its mechanism must not only function properly but the hands must be correctly set beforehand. The analogy can be considered either purely mechanistically or else mathematically. In the latter case the object is to calculate the course which a physical system will follow in time from given initial conditions.

This was the method adopted by Newton in the theory of gravitation which he developed in the *Principia*, published in 1687, the full title of which refers specifically to the *mathematical* principles of natural philosophy. As he said in one of his letters to Bentley, 'Gravity must be caused by an agent acting constantly according to certain laws; but whether this agent be material or immaterial I have left to the consideration of my readers.' Unlike his principal continental critics, Huygens and Leibniz, Newton was willing, at least in the *Principia*, to bypass the problem of explaining gravitation mechanistically. Instead, taking time as the independent variable, he formulated mathematical laws of motion and gravitation in terms of which gravitational phenomena can be described and predicted.

The particular concept of mathematical time used by Newton was based on the analogy between time and a geometrical straight line. Although this analogy had been used by Galileo and by others before him, notably Nicole Oresme in the fourteenth century, the first explicit account of it was given by Isaac Newton's predecessor in the chair of mathematics at Cambridge, Isaac Barrow, in his *Geometrical Lectures*, published in 1670. Barrow was greatly impressed by the kinematic method in geometry that had been developed by Galileo's pupil Torricelli. Barrow realized that to understand this method it was necessary to study time, and he was particularly concerned with the relation of time and motion. 'Time does not imply motion, so far as its absolute and intrinsic nature is concerned; not any more than it implies rest; whether things move or are still, whether we sleep or wake, Time pursues the even tenour of its way.' He regarded time as essentially a mathematical concept that has many analogies with a line, for 'Time has length alone, is similar in all its parts and can be looked upon as constituted from a simple addition of successive instants or a continuous flow of one instant.'[12] Barrow's statement goes further than any of Galileo's, for Galileo used only straight line segments to denote particular intervals of time.

Barrow's views greatly influenced Newton. In particular, Barrow's idea that, irrespective of whether things move or are still, whether we sleep or wake, 'Time pursues the even tenour of its way' is echoed in the famous definition at the beginning of Newton's *Principia*: 'Absolute, true and mathematical time, of itself, and from its own nature, flows equably without relation to anything external.' Newton regarded the moments of absolute time as forming a continuous sequence like the points on a geometrical line and he believed that the rate at which these moments succeed each other is independent of all particular events and processes.

Newton's adoption of the idea of. absolute time, existing in its own right, was partly due to his belief that there must be a fundamental theoretical measure of time to compensate for the difficulty of determining a truly accurate practical time-scale. As has been discovered since (see pp. 167–8) and as Newton himself seems to have realized, in the long run we cannot obtain a truly fundamental time-scale from the observed motions of either the earth or the heavenly bodies. One of the difficulties of Newton's definition, however, is that there is no way of using it to obtain a practical means of measuring time. It has also been criticized by philosophers, since it ascribes to time the function of flowing; but, if time were something that flowed, then it would itself consist of a series of events in time, and this is meaningless. *Time cannot itself be a process in time.* Moreover, what is meant by saying that 'time flows equably' or uniformly? This would seem to imply that there is something which controls the rate of flow of time so that it always goes at the same speed. If, however, time exists 'without relation to anything external', what meaning can be attached to saying that its rate of flow is not uniform? If no meaning can be attached even to the possibility of non-uniform flow, what is the point of stipulating that its flow is 'equable'?

That moments of absolute time can exist in their own right is now generally regarded by scientists and philosophers as an unnecessary hypothesis. Events are simultaneous not because they occupy the same moment of time but because they occur together. Any two events that are not simultaneous are in a definite temporal order since one occurs before the other and not because they occupy different moments of time, one of which is earlier than the other. In other words, we derive time from events and not the other way round. This was the point of view taken by Newton's great contemporary Leibniz, who did not believe that moments of time can exist independently of events. He based his argument for this on what he called 'the principle of sufficient reason', according to which nothing happens without there being a reason why it should be so rather than otherwise. He applied this principle to time by considering the case of someone who asks why God did not create everything a year sooner and from this wishes to infer that God did something for which he could not possibly have had a reason to do it thus rather than otherwise. Leibniz's reply is that the inference would be true if time were something apart from temporal things, for it would be impossible that there should be reasons why things should have been applied to certain instants rather than to others when their succession remained the same. But this itself proves that instants apart from things are nothing and that they only consist in the successive order of things. If this remains the same, one of the states (for example, that in which the Creation was imagined to have occurred a year earlier) would be in no way different and could not be distinguished from the other. In Leibniz's view, *time is the order of succession of phenomena*, so that if there were no phenomena there would be no time.[13]

The nature of time and its relationship to different forms of existence, including the physical, had been considered long before the seventeenth century, notably by St Thomas Aquinas (1224–74) in his massive *Summa theologica*, in which he discussed three kinds of 'time'. Time, in the strict sense, he regarded as a state of succession that has a definite beginning and end. It applies only to terrestrial bodies and phenomena. Eternity, which exists all at once (*tota simul*), is essentially 'timeless' and the prerogative of God alone. The third concept, called *aevum*, originally due to the sixth-century philosopher Boethius, like time has a beginning but unlike time no end. Aquinas considered it to be the 'temporal' state of angels, heavenly bodies, and ideas (*archetypum mundum*).

Despite the difference between Newton's and Leibniz's views concerning the nature of physical time, in other respects their ideas about the concept were similar. Both believed that time was universal and unique, the universe comprising a succession of states, each of which exists for an instant, successive instants being like the order of points on an indefinitely extended straight line. This was the concept of time which was to dominate physical science until the advent of Einstein's special theory of relativity early this century.

Newton's views concerning time were not confined to the physical world but extended to human history and to prophecy. Like many of his contemporaries he believed that the world was coming to an end. He was convinced that the comet of 1680 had just missed hitting the earth, and in his biblical commentaries, on Revelations and the Book of Daniel he maintained that the end of the world could not long be delayed, but he was careful to avoid the prediction made by the millenarians who had settled upon a date. His contemporary and fellow scientist Robert Boyle also believed that 'the present course of nature shall not last always, but that one day this world . . . shall either be abolished by annihilation, or, which seems more probable, be innovated, and, as it were transfigured, and that, by the intervention of that fire, which shall dissolve and destroy the present frame of nature'.[14]

Newton's *Chronology of Ancient Kingdoms amended*, posthumously published in 1728, and his *Observations upon the Prophecies of Daniel and the Apocalypse of St John*, published in 1733, can together be regarded as providing a universal history of mankind that was intended to be the counterpart of the physical history of the world set out in his *Principia*. By about 1700 chronology had become a subject of major concern to many thinkers because of its relevance for the authenticity of the Bible. The Old Testament as it has come down to us contains no dates. Bede had calculated the interval between Creation and Incarnation to be 3,952 years. Earlier, Eusebius obtained a figure of 5,198 years. By 1660 at least fifty different dates had been assigned to Creation, depending on which version of the Old Testament and which counting method were used.[15] James Ussher, Archbishop of Armagh (1581–1656), proposed 23 October 4004 BC, and the Danzig astronomer Johannes Hevelius (1611–87) in his *Prodromus astronomiae*, posthumously published in 1690, computed the exact time to be 6 o'clock in the evening, 24 October 3963 BC.[16] Newton, however, was careful not to assign a specific date for the Creation.

Newton devoted a good part of the last thirty years of his life to the careful study of chronology and sought to determine what he regarded as key dates, such as that of the Argonauts' expedition. Although he made use of literary references whenever necessary, he preferred to use astronomical techniques if possible. In particular, he thought that chronology could be put on a scientific basis by means of the accurate determination of the precession of the equinoxes. By its aid he believed that, if a relevant record could be found of the position of the sun relative to the fixed stars at the time of equinox, any event in the past could, in principle, be dated.

In a letter to Oldenburg of 7 December 1675, Newton explicitly stated his belief that 'nature is a perpetual circulatory worker'. Although later he argued that the 'amount of motion' in the world would of its own accord tend to decrease, *unless God intervened to correct this*, this proviso reveals his continuing belief in the essentially clocklike nature of the universe. In his view, God actually needed to intervene in the natural world from time to time to adjust its working in the same kind of way as a clockmaker needs occasionally to reset a clock so that it reads correctly.

Attitudes to time and history in the sixteenth and seventeenth centuries

In the sixteenth century people tended to be obsessed with the destructive aspect of time. The typical Renaissance image of time was as the destroyer equipped with hour-glass, scythe, or sickle. This attitude to time can be seen in Shakespeare, notably in his sonnets and in *The Rape of Lucrece*, especially stanza 133:

> Mis-shapen Time, copesmate of ugly Night,
> Swift subtle post, carrier of grisly care,
> Eater of youth, false slave to false delight,
> Base watch of woes, sin's pack-horse, virtue's snare,
> Thou nursest all and murder'st all that are:
> O, hear me then, injurious, shifting Time!
> Be guilty of my death, since of my crime.

In Shakespeare's sonnets time is treated with what has been described as a 'polyphonic grandeur unmatched in English literature'. In some ways his attitude to time appears to have been very different from ours. For example, whereas we like to think of his plays as having been written for all time, it is most unlikely that that is how he thought of them himself. In his day the average run of a play was not more than five performances, few plays were ever revived and hardly any were printed. It is probable that Shakespeare wrote his plays not for the sake of posterity, but to earn enough money for him to be able to retire in comfort to his native Stratford-upon-Avon. A distinguished Tudor historian has pointed out that when Shakespeare lived, 'No playwright in his senses could have supposed himself to be writing for all time. The Elizabethans lived in the present.'[17] To them *ars longa, vita brevis* would have been meaningless.

While Shakespeare's concern with time was only at the personal level, his contemporary Edmund Spenser was obsessed with time at all levels, including the astronomical.[18] Although the Church Fathers had converted history from an endless sequence of cycles to a vision of the whole universe moving from Creation to Redemption, the figure of the circle still dominated human thought in astronomy in the sixteenth century. This had a great influence on Spenser who, despite his profound concern with time, was essentially a backward-looking figure. As a recent authority on the role of time in his poetry has remarked, Spenser believed that 'our mortality and the insufficiency of all created things is, by grace, only one aspect of a total situation of which cyclical return is the other face, until such time as time shall cease'.[19]

Another contemporary of Shakespeare, Sir Walter Raleigh, was also greatly concerned with the depredations of time. He believed that the objective order of the universe was revealed in history and that it provided a vision of the meaning and purpose of human life. His massive *History of the World*, composed between 1608 and 1614 during his imprisonment in the Tower of London, was dedicated to James I, but instead of being pleased that learned but tetchy monarch complained that the book was 'too sawcie in censuring princes'! It was dominated by Raleigh's preoccupation with time, especially the poignant contrast between the temporal scale of his own life and the vast enterprise he had undertaken. Although he left his book incomplete after covering the period from the Fall of Adam to the Fall of Carthage, it runs to over 2,700 pages in the reprint of 1829. Well aware of the cosmic significance of time, he was convinced that all along the world was tending to get worse and worse. In holding this belief he was in general accord with the prevailing opinion of thinkers and writers of the Renaissance and Reformation eras, who were almost entirely backward-looking. Overwhelmed by a sense of the significance of the 'Cosmic Fall', they tended to believe in the existence of a primeval 'Golden Age', followed by irreversible decline. One of the starkest expressions of this view was Martin Luther's in his commentary of 1545 on the Book of Genesis: 'The world degenerates and grows worse every day. . . . The calamities inflicted on Adam . . . were light in comparison with those inflicted on us.' Luther also complained that after the Flood the trees and fruits of the earth 'are but miserable remnants . . . of those former riches which the earth produced when first created'.[20]

The backward-looking tendency in the sixteenth century is indicated by the word *Rinascita* ('Renaissance'), invented by Vasari and others in Italy, for it signified the rebirth of something old and not the introduction of something new. Later that century rulers such as Philip II of Spain, Elizabeth I of England, and Henry IV of France thought of themselves as upholders and maintainers and never as founders and innovators. Indeed, as a biographer of the French monarch has pointed out, 'Such an attitude caused Henry to describe himself to the Assembly of Notables as "Liberator and Restorer of the French State".'[21] These rulers regarded their reforms as a return to some pristine model of the past.

Similarly, Vieta (François Viète), the greatest mathematician of the sixteenth century, regarded innovation as renovation. Moreover, even the numerous technological advances in western Europe in the Middle Ages led to no general concept of technological progress.

During the Renaissance men became more and more aware that almost everything changes with time and so has a history. However, whereas in the Middle Ages the linear interpretation of history had been stressed because of its significance for Christian doctrine, in the Renaissance there was a marked revival of the cyclical view, because there was more concentration on secular history. Greater attention was paid to the literature that had survived from classical antiquity and to the cyclical point of view that characterized much of it. For example, Giorgio Vasari (1511–64) in his *Lives of the Painters, Sculptors, and Architects* favoured this idea in the history of art, believing—not surprisingly—that after Michelangelo (1475–1564) art was more likely to decline than to progress still further. More surprisingly, Francis Bacon (1561–1626), the prophet of scientific advance, in one of the last essays that he wrote, 'Of Vicissitude of Things', adhered to a cyclical view of history in general:

In the *Youth* of a *State, Armes* doe flourish: in the *Middle Age* of a *State, Learning*: And in them both together for a time: In the *Declining Age* of a *State, Mechanical Arts* and *Merchandise. Learning* hath his Infancy, when it is but beginning, and almost Childish: Then his Youth, when it is Luxuriant and Iuventile: Then his Strength of yeares, when it is Solide and Reduced: and lastly his Old Age, when it waxeth Dry and Exhaust. But it is not good, to looke too long, upon these turning Wheeles of *Vicissitude*, lest we become Giddy.

In the course of the seventeenth century the pessimistic and backward-looking attitudes to time that had characterized the previous century were gradually replaced by optimistic and forward-looking views. An optimistic view of the future was expressed by Francis Bacon in an early unpublished essay of 1603. It bore the significant title *Temporis partus masculus (The masculine birth of time).*

Mary Tiles in a recent paper with the title 'Mathesis and the Masculine Birth of Time' has discussed Bacon's ideas on scientific method and his peculiar terminology.[22] Something is a 'birth of time' if it arises from cumulative corporate experience. Truth was regarded by Bacon as the 'feminine birth of time', whereas by the 'masculine birth of time' he meant active intervention in the world amounting to an exercise of power over nature. Bacon distinguished knowledge derived from ancient texts from that actively sought by modern natural philosophers. The term 'mathesis' refers to the ordering of knowledge (classification, etc.), particularly by means of mathematics. Bacon wrote: 'Science is to be sought from the light of nature, not from the darkness of antiquity. It matters not what has been done; our business is to see what can be done.'

The scornful rejection by Bacon of the doctrine that the ancients had encompassed all knowledge was echoed by, among others, John Wilkins who in *The Discovery of a New World*, published in 1638, attempted to show that the moon is inhabited. In it he wrote: 'There are yet many secret Truths which the ancients have passed over, that are yet left to make some of our Age famous for their Discovery.' Two years later, in *A Discourse concerning a New Planet*, in which he advocated the Copernican theory, he wrote even more significantly:

Antiquity does consist in the Old Age of the World, not in the youth of it. In such learning as may be increased by fresh Experiments and new Discoveries: 'tis we are the Fathers, and of more Antiquity than former Ages; because we have the advantage of more Time than they had, and Truth (we say) is the Daughter of Time.

This slogan (*Veritas filia temporis*) was much used in the sixteenth century and has a fascinating history, as has been shown by Fritz Saxl in his erudite chapter bearing this title in the *Ernst Cassirer Festschrift*.[23] In an important footnote (p. 200) he mentions that his 'learned friend' Dr Klibansky had informed him that it can be traced back to ancient Greece where two different traditions prevailed: 'Time reveals either guilt and its punishment, as in Aeschylus' tragedies, or it reveals true valour and the honour due to it, as in Pindar's aristocratic poetry. Sophocles uses it to express his humble faith in the justice of divinity.'

The change from a backward to a forward temporal orientation was also advocated, in 1627, by a young Anglican divine, George Hakewill, in a refutation of Bishop Goodman's *Fall of Man* (1616), an unremittingly gloomy demonstration that the world was approaching extinction. The question of the 'last days' aroused the intellectual interests not only of chronologists but also of mathematicians. Baron Napier of Merchiston, who published his famous *Mirifici logarithmorum canonis descriptis* in 1614, valued his invention of logarithms because it helped to speed up his calculations of the number of the Beast in the Book of Revelations, whom he wished to identify with the Roman Pope! Hakewill's book of over 600 pages bore a lengthy title that began: *An*

Apologie of the Power and Providence of God in the Government of the World or an Examination and Censure of the Common Errour Touching Nature's Perpetuall and Universall Decay . . . In this prolix publication Hakewill, who was influenced by Bacon, dismissed the traditional lamentations about decay as merely a manifestation of 'the morosity and crooked disposition of old men, always complaining of the hardness of the present times, together with an excessive admiration of Antiquity'.[24] At first his views met with considerable opposition, but as the century progressed they came to be widely accepted. For example, we find Milton declaring in one of his Latin essays that 'Natura non patet senium' ('nature does not suffer from old age').

The world was not necessarily getting worse; still less did it show signs of ending, despite the prophets of the 'Fifth Monarchy', who had announced that the Millenium would arrive in 1666. Why believe Nostradamus, who a century earlier, had assigned the end of the world to 31 July 1999? An important contributory factor to the development of a sense of time associated with a forward-looking perspective of political action was provided by the 'Antichrist' myth which became widespread in England in the mid-seventeenth century during the religious and political upheavals of the civil war and its aftermath. This myth looked forward to the defeat of Antichrist, variously identified with the Pope, the bishops, the whole hierarchy of the Church of England, the King, and the royalists. Gradually, however, as Doomsday was repeatedly postponed, the Golden Age was transferred from the past to the future and millenarian prophecies were replaced by Utopian programmes. As Carl Becker has neatly summarized it, in the eighteenth century 'The philosophers called in Posterity to exorcise the double illusion of the Christian paradise and the golden age of antiquity.'[25]

A forward-looking perspective also greatly influenced those who rejected scholastic philosophy and replaced it by the experimental philosophy advocated by Francis Bacon. The scientific revolution of the seventeenth century gave rise to what has been called 'the quarrel of ancients and moderns'. The essential point at issue was that unquestionable authority should no longer be attributed to the thinkers and writers of antiquity. In France a general attack on appeals to authority in scientific matters was made by Bernard de Fontenelle in his *Digression sur les anciens et les modernes*, published in 1683.

The need to establish an objective criterion of historical truth was clearly realized by Jean Bodin in his *Methodus ad facilem historiarum cognitionem* (*Method for the easy understanding of histories*) of 1566, but although he castigated those who sighed for a lost Golden Age, he still adhered to the cyclical concept of history, as did the Italian philosophers of history Machiavelli and Guiccardini earlier that century. Machiavelli thought that history is dominated by an oscillation between the bad and the good but with the bad tending to be in control longer than the good. Belief in the cyclical nature of history was, in fact, widespread from the Middle Ages to the seventeenth century despite the Church's view that time extends from the Creation to the end of the world, both being unique events. Despite his cyclical views, Bodin was one of the first to attempt to discover whether there are any causal factors controlling the rise and fall of empires and so producing a common direction of historical events. According to a modern authority, he also gave 'one of the best early surveys of the history of historiography'.[26]

Although the Bible provides no dates, its chronology, particularly that of the Old Testament, became important following the Reformation and the resulting theological disputes. Previously, the Bible had not been regarded by the Church as a historical document but allegorically as an oracle. The view of Protestants such as Luther, however, was that the Bible (which for them replaced the Church as the ultimate source of religious authority) should be taken literally—a point of view that has not been entirely eliminated even in our own day. In England, Richard Hooker (c.1554–1600), in the course of his intellectual defence of the theological and ecclesiastical *media via* of the Anglican Church, *The Laws of Ecclesiastical Polity*, criticized the Puritans for attempting to apply Old Testament precepts to contemporary society, for which they were irrelevant because of the very different circumstances that prevailed. In the following century, the philosopher Baruch Spinoza (1632–77) went still further in regarding the Bible purely as an historical document and so was a forerunner of the nineteenth- and twentieth-century historical experts in the higher criticism of that work.

Scholarly historical chronology, in particular of classical antiquity, began with the publication in 1583 of the *De emendatione temporum* of the great scholar J. J. Scaliger (1540–1609). He introduced in 1582 the system of Julian days beginning at noon on 1 January 4713 BC (for chronological purposes this was the date chosen by him for the Creation) so as to avoid the irregularities in length of the months and years when calculating the time between two events. The number of the Julian day beginning at noon on 1 January 1988 is 2,447,162. Julian days are still used by astronomers, for example, for the times of maximum and minimum brightness of variable stars.

Despite Scaliger, when the philosopher René Descartes (1596–1650) was engaged in his quest for absolute certainty he dismissed history as being based on mere opinion and arbitrary subjectivity, the historical sciences in his day being in a more primitive state than the mathematical. Indeed, the first formulation of criteria for testing the authenticity of documents, particularly charters and other manuscripts in medieval Latin, was not achieved until after Descartes' day by Mabillon in his *De re diplomatica*, published in 1681. The process of corruption of texts was ultimately arrested by printers, but not until the eighteenth century. At first, following the invention of movable type about 1450, texts actually appear to have been altered more rapidly by early printing methods than they had been by medieval copyists.[27]

Another work published in 1681 was the first French detailed world-history, Jacques Bossuet's *Discours sur l' histoire universelle*; but although he dealt with the rise and fall of empires, Bossuet omitted all those that were not Christian, except for Greece and Rome in so far as he considered that they were relevant for the establishment of Christianity. Nevertheless, Bossuet's book is important in the history of historiography because it was one of the first universal histories after Raleigh's. Bossuet believed that man's actions are supervised by a Divine Providence, so that, however inexplicable and surprising particular events may seem to be, they nevertheless advance in '*une suite reglée*'.

9. Time and History in the Eighteenth Century

The invention of the marine chronometer

In the eighteenth century the outstanding achievement concerning time was the invention of the marine chronometer, which revolutionized navigation and thereby saved countless lives. The practical need for precise timekeeping at sea dates from the voyages planned by Prince Henry the Navigator in the fifteenth century. After the Cape of Good Hope had been rounded in 1488 it was east–west distances that mattered, and the need arose to determine longitude at sea. This was more difficult than determining a ship's latitude, since that could be obtained by measuring the altitude of the sun at local apparent noon with the aid of the cross-staff or the astrolabe. (The sextant did not come into practical use until later.) Whereas the poles of the earth's axis of rotation serve as universal reference points for the measurement of latitude, there are no such natural aids for the determination of longitude. Instead, an arbitrary zero of longitude has to be chosen. This is called the 'prime meridian'. The longitude of a place can then be obtained by determining the time that would be taken by the earth to turn through the angle which would bring the meridian through that place into the position the prime meridian was in at the epoch when the ship's position was being determined. The definition of longitudinal distances on land by means of differences in local time was known to the Greeks of the third century BC and possibly earlier.

Two different ways of solving the problem of a ship's longitude were suggested. One was astronomical, being based on observing the position of the moon relative to the stars. This method was proposed in 1514 by Johann Werner of Nuremberg (1468–1522) and became known as the lunar-distance method. The other, first suggested in 1553 by Gemma Frisius of Louvain (1508–55), depended on the development of an accurate chronometer that would be designed so as to withstand the disturbances associated with transport by sea. It would be set to give the time on the prime meridian, and this would be compared with the local time (e.g. noon) of the place where the ship was situated.

In 1567 Philip II of Spain offered a substantial reward for a successful way of determining longitude at sea, and this reward was greatly increased thirty years later. Among those applying for the prize was Galileo, who realized that the discovery he had made with the aid of his telescope in 1610, of the four principal satellites of Jupiter and their occultations by that planet might be the basis of an accurate celestial timekeeper. He submitted his method in 1616, but the Spaniards did not regard it as a practical proposition. Galileo also contributed to the chronometric method of solving the problem by drawing attention to the possibility of using the pendulum as a controller of mechanical clocks. Later that century, as we have seen, the pendulum clock was successfully developed by Huygens, but although he was convinced that his clocks could be used to determine accurate longitudes they tended to be erratic except on land or on a calm sea.

Meanwhile the lunar-distance method had been revived in Paris by J.-B. Morin (1583–1656), who suggested that an observatory was needed to provide the required data. Following the creation in 1666 of the Académie Royale des Sciences by Louis XIV at the instigation of the great statesman Colbert, the Paris Observatory was founded a year later. 'Finding the longitude' was also one of the subjects that engaged the Royal Society of London, established by Charles II in 1660, and led to the founding of the Royal Observatory at Greenwich in 1675.

Any astronomical method for finding differences of longitude must assume that for practical purposes the earth rotates uniformly. Since solar days vary throughout the year, the first Astronomer Royal, John Flamsteed (1646–1719), decided to check this assumption by concentrating on sidereal time. With the aid of two clocks installed in 1676 in the Octagon Room in the Royal Observatory at Greenwich he concluded that the earth does rotate uniformly, a result which was not challenged for 250 years. These clocks had been made by Thomas Tompion (1639–1713), who has been called 'The father of English clockmaking'.[1] Like some of the earlier English clockmakers he began as a blacksmith. In due course he became a friend of Robert Hooke (1635–1704), Curator of Experiments at the Royal Society, and through him met Flamsteed. The two clocks he made for Flamsteed could each run automatically for a year.

Although the expression 'finding the longitude' was often used as a catch-phrase like 'squaring the circle' to denote something thought to be impossible, the need to solve the problem became acute after a spectacular maritime disaster on 29 September 1707 when Admiral Sir Clowdisley Shovel and some 2,000 other sailors were drowned because bad navigation had caused four ships of the Royal Navy, on the way home from Gibraltar, to capsize on the Gilstone Ledges in the Scilly Isles. This disaster led to a public outcry for navigation to be made more accurate. 'Finding the longitude' seemed to be the key to this, as was made clear by Sir Isaac Newton when he appeared before a Parliamentary Committee that had been set up to examine the problem:

That, for determining the Longitude at Sea, there have been several Projects, true in theory, but difficult to execute. One is a Watch to keep Time exactly, but, by reason of the motion of the Ship at Sea, the Variation of Heat and Cold, Wet and Dry, and the Difference of Gravity in different latitudes, such a Watch has not yet been made.[2]

As a result, on 8 July 1714 Queen Anne gave the Royal Assent to *A Bill for Providing a Publick Reward for such Person or Persons as shall Discover the Longitude at Sea*. A prize of £20,000 was offered, equivalent to over £1 million today, for a method of determining the longitude at sea to within thirty geographical miles at the end of a voyage to the West Indies. One 'geographical mile' is equivalent to one arc-minute of longitude at the Equator (6,087 ft). Since one degree of longitude corresponds to 4 minutes of time, award of the full prize meant that the chronometer had to be accurate to within 2 minutes after about 6 weeks of sailing. Smaller prizes were offered for accuracy to within 40 miles (£15,000) and sixty miles (£10,000). The power to make the award was vested in a special Board of Longitude, the Commissioners of which were twenty-two sailors, politicians, and scholarly experts who were answerable direct to Parliament.

For years this government Act did little to diminish general scepticism about the possibility of solving the problem. Jonathan Swift in the 'Voyage to Laputa', the third of *Gulliver's Travels*, published in 1726, remarked that only if Gulliver became immortal like the Struldbrugs would he 'then see the discovery of the *longitude*, the *perpetual motion*, the *universal medium*, and many other great inventions brought to the utmost perfection'.[3] Nine years later the painter Hogarth went even further by including in the final madhouse scene of 'The Rake's Progress' a man who is trying to calculate the longitude! Despite the large prize offered, over twenty years elapsed before the Commissioners had anything to record in their minutes, so great appeared to be the difficulties in arriving at a satisfactory solution. Nevertheless, the practical need to solve the problem became ever more pressing. In April 1741 after Commodore (later Admiral) Anson in the *Centurion*, accompanied by other vessels, had rounded Cape Horn, an unsuspected easterly current prevented him from travelling as far westward as he thought. Since many of his men were dying from scurvy, he was anxious to land on the island of Juan Fernandez for fresh vegetables, but the double uncertainty concerning the longitude of his ships and that of the island (a consequence of the inability of explorers to determine the longitude of their discoveries) meant that it was nearly the middle of June before he had arrived there. This delay cost the lives of nearly seventy of his men.

By an ironical coincidence, some five years previously the *Centurion* had been the first ship in the history of navigation to carry, on a trial voyage to Lisbon, a clock which it was thought might provide a practical means of determining longitude at sea. The clock used in this test had a special kind of balance-spring, since it had been realized for more than half a century that pendulum clocks were not suitable for use at sea, because of the effects of pitching and rolling. Balance-springs, however, were found to be particularly sensitive to changes of temperature, losing time in hot weather and gaining it in cold. In the case of the pendulum clock the first horologist to overcome a similar difficulty was Tompion's former assistant George Graham (1673–1751). In 1726 he used mercury to counteract the expansion and contraction of the pendulum. It was arranged that, for example for a rise in temperature, the upward expansion of mercury in the bob would counteract the downward expansion of the steel pendulum rod, so that the period of swing of the pendulum was unaltered. Already, in 1715, he had invented an improved form of escapement, known as the dead-beat escapement. Although Graham's regulator clock of 1730 which involved both these devices proved to be an excellent timekeeper on land, it did not solve the problem of determining longitude at sea.

The honour of solving that problem fell to John Harrison (1693–1776), who was originally a carpenter in Yorkshire. With his brother James he first of all produced, in 1728, a clock with a pendulum made of brass and steel rods so arranged that it was practically temperature-independent. This clock also had a complicated escapement involving a minimum of friction, which had been another source of inaccuracy in clocks. The Harrisons then went on to invent an accurate chronometer for use at sea. By 1735 it was completed and the following year, on the recommendation of the Royal Society, it was tried out successfully on a voyage of the *Centurion* to Lisbon. This chronometer involved two straight-bar balances, with a ball at each end, which were pivoted at their centres and were connected by helical springs and cross-wires so as to swing as if geared together, but with far less friction. It was found that the motion of a ship had very little effect on their period of oscillation. As in the Harrisons' earlier clock, an ingenious combination of brass and steel rods varied the tension in the springs so as to compensate for the effect of changes of temperature. This device was the first system of compensation for temperature applied to a balance-clock, and the machine itself was the first accurate marine chronometer.

Following the successful trial of this chronometer on the voyage to Lisbon, the Board of Longitude met on 24 June 1737. John Harrison was present, but instead of asking that his chronometer should be tested on a voyage to the West Indies he offered to make an improved version for that purpose. The Board of Longitude resolved to advance him £500. Harrison's second chronometer, however, was never given a trial at sea, presumably because war with Spain had broken out and so it would have been exposed to the danger of capture. It is also possible that Harrison may have had some doubts about its performance.[4] Be that as it may, another seventeen years were to elapse while he concentrated on producing a third clock, which he intended to be his masterpiece. Throughout these years the Royal Society, influenced by Graham, helped to support him and in 1749 bestowed on him its highest award, the Copley Medal.

Harrison's third chronometer was the most complicated of all his machines, containing no fewer than 753 separate parts. At last, in 1757, he notified the Board of Longitude that he proposed to compete with it for the £20,000 prize. At the same time he offered to make a much smaller timekeeper to serve as an auxiliary. This proposal was accepted, and aided by his son William John Harrison constructed his famous 'watch', which on test he found to be as accurate a timekeeper as his third chronometer, while possessing the advantage of being much more portable. It was a large silver watch just over 5 in. in diameter. In outward appearance it resembled the ordinary 'carriage-watch' of the period, but in essentials it was similar to his third chronometer except for the escapement, which was a greatly improved version of the usual verge watch-escapement of the period. (A detailed description of this famous chronometer has been given by Gould.[5]) Because of its accuracy and portability Harrison decided to compete with this fourth chronometer alone. Consequently, it was officially submitted for trial on the journey to Jamaica in 1762. It easily passed the test, being only 5 seconds slow on arrival there, corresponding to one and a quarter arc-minutes of longitude, which in the latitude of Jamaica was less than one geographical mile. Harrison, therefore, expected to receive the £20,000 prize. Instead, the Board of Longitude allowed him only £2,500 on account, because in their opinion the longitude of Jamaica was not known accurately enough to provide a sufficiently precise standard of time!

Meanwhile others were competing for awards from the Board of Longitude. Because of the complicated nature of the moon's motion (it was the only problem that Newton said ever made his head ache!), the lunar-distance method of determining longitude was not of practical use until the German astronomer Tobias Mayer (1723–62) produced his tables of the moon's motion with the aid of calculations that had been made by the great mathematician Leonhard Euler (1707–83). Mayer submitted his application for the Board of Longitude's prize in 1755. Ten years later his widow was awarded £3,000 in recognition of his achievement and £500 pounds was awarded to Euler.

After much argument and delay over Harrison's prize a second official trial of his fourth chronometer was undertaken some two years after the first, on a voyage to Barbados. Because of his age, John Harrison did not travel on either voyage but his son William went instead. On the second voyage he was accompanied by two astronomers, one of whom was Nevil Maskelyne, who shortly afterwards became Astronomer Royal. They were instructed to determine the longitude of their observation-point on Barbados astronomically. By this time the new lunar tables, due to Euler and Tobias Mayer, were available, and the sextant had replaced the clumsier quadrant. Consequently, celestial observations were not only easier to make but could be more accurately determined and readily checked.

Early in 1765 it was reported to the Board of Longitude that the chronometer's error was only 38.4 seconds (in seven weeks) corresponding to 9.8 geographical miles at the latitude of Barbados.[6] Although this result was three times better than required, the Board remained sceptical and refused to pay Harrison until he had made a full disclosure on oath of the details of the mechanism of his chronometer. The Board would then pay him £10,000, less the £2,500 already paid to him in 1762 after the Jamaica trial. The remaining £10,000 they refused to pay until he had made two more successful timekeepers. Eventually, Harrison accepted the first half of the reward. By 1770 he had made, with his son's help, a fifth timekeeper which was a slightly improved version of the fourth.

Meanwhile King George III's interest had been aroused, and at an audience at Windsor granted to Harrison and his son he exclaimed, 'By God, Harrison, I will see you righted!'[7] The fifth chronometer was then tested at the King's private observatory at Kew. In ten weeks its total error was only four and a half seconds. Nevertheless, the Board of Longitude objected that this trial had not been authorized by them and they refused to accept its results. Harrison then petitioned the House of Commons, where his case was supported by Fox and Burke, among others. As a result, the Board finally decided to pay Harrison another £8,750, arguing that the remaining £1,250 pounds had already been paid to him many years earlier on the understanding that his second and third chronometers should become the property of the Board! He died three years later.

Harrison's achievement proved to be a landmark in the history of time measurement. Only after the success of the Jamaica trial in 1762 did most makers of clocks and watches begin to realize that high-precision timekeeping at sea was a practical possibility. Harrison's success also had a tremendous influence on the construction of maritime charts. In the second of Captain Cook's voyages of exploration in the South Pacific between 1772 and 1775 an exact replica of Harrison's fourth chronometer enabled him to construct maps of the coastlines of Australia and New Zealand of great accuracy.

Outstanding among those in other countries in the eighteenth century who contributed to the development of the marine chronometer was Pierre Le Roy (1717–85). In 1754 he succeeded his father in the post of 'Horloger du Roi' to Louis XV of France. By then he had invented an improved form of escapement known as the 'detached escapement', in which the motion of the balance was effectively a free vibration, subject only to minimal disturbance at the instants of receiving impulse and actuating the escapement. Later Le Roy invented the 'compensation balance', which corrected the effect of change of temperature in altering the elasticity of the spring and the moment of inertia of the balance. Although Harrison was indisputably the first man to make a satisfactory marine timekeeper, the modern chronometer owes more to the inventions of Le Roy.[8] Contemporary with Le Roy was his rival Ferdinand Berthoud (1729–1807), who was born in Switzerland but spent nearly all his working life in France. Although Berthoud did not have Le Roy's profound understanding of the basic principles of chronometry, he was like Harrison a superb craftsman who continually made improvements to his machines. Nevertheless, they remained complicated and expensive.

Meanwhile in England there was a concerted move to simplify marine chronometers and make them cheap enough for the ordinary navigator. The two principal pioneers in this important development were John Arnold (1736–99) and Thomas Earnshaw (1749–1829). In particular, Earnshaw improved the compensation balance that had been invented by Le Roy. As a result, from about 1825 marine chronometers became standard equipment in all ships of the Royal Navy, the East India Company having already insisted on this in their ships some years previously.

The discovery of historical perspective

The eighteenth century was important in the history of time not only because of the invention of the marine chronometer, but also because the spirit of intellectual optimism that characterized the age of enlightenment, as that century came to be called, was based on a forward-looking attitude to time. The thinker who has been particularly associated with the emergence of this point of view is Leibniz, who maintained that this is 'the best of all possible worlds'. Although he has often been derided for this claim, which indeed took a severe knock from the great Lisbon earthquake of 1755, in fairness to him the emphasis should be placed on the word 'possible', for he did not believe that this world is actually 'perfect'. R. Nisbet's recent book on the history of the idea of progress has drawn attention to the following passage in Leibniz's essay 'On the Ultimate Origination of Things' which makes this quite clear.

To realize in its completeness the universal beauty and perfection of the works of God, we must recognize a certain perpetual and very free progress of the whole universe such that it is always going forward to greater improvement. . . .

And to the possible objection that, if this were so, the world ought long ago to have become a paradise, there is a ready answer. Although many substances have already attained a great perfection, yet on account of the infinite divisibility of the continuous, there always remain in the abyss of things slumbering parts which have yet to be awakened, to grow in size and worth, and in a word, to advance to a more perfect state. And hence no end of progress is ever reached.[9]

Arthur Lovejoy has pointed out that one of the principal features of eighteenth-century thought was the temporalizing of what has been called 'the great chain of being', that is the idea that the universe is composed of an immense number of links arranged in hierarchical order. For, although to many minds of that century the idea of a world in which no emergence of novelty was possible seemed completely satisfying, there were others who felt that the chain of being 'must perforce be reinterpreted so as to admit of progress in general'.[10]

Among those in the eighteenth century who philosophized about the historical process as a whole and tried to discover the laws that govern it were Turgot and Condorcet in France, Priestley in England, and Kant in Germany. It has been claimed that Turgot's address at the Sorbonne in December 1750, when he was only 23 years old, on 'A Philosophical Review of the Successive Advances of the Human Mind' was the first systematic statement of the modern idea of progress.[11] In his *Notes on Universal History*, which he wrote the following year, he acknowledged the influence of Bossuet, but his approach to the subject was purely secular. In it he showed that, while the natural talents of men are everywhere the same, the particular characteristics of a society are the inevitable consequences of its own past. Turgot's writings greatly influenced Condorcet, who wrote the first biography of Turgot, whom he regarded as the real discoverer of the 'law of progress'. Like so many of his contemporaries, including the young Wordsworth, Condorcet was convinced of his good fortune in living during one of the greatest revolutions in history and being able to recognize its true significance. Ironically, he came under suspicion following the expulsion from the Convention in June 1793 of the Girondins, with whom he was friendly. He managed to hide in a house in Paris, where he composed his *Sketch*, but on 25 March 1794 he left the house and two days later was arrested. The following morning he was found dead in his cell.

In his *Sketch for a Historical Picture of the Progress of the Human Mind*, published in 1795, one year after his death, Condorcet expressed his belief in the inevitability of human progress and in the power of science and technology to transform man's knowledge and control over himself and society. He viewed history as a sequence of ten distinct stages, each arising necessarily from the preceding one. In the first stage man lived in a state of primitive savagery. In the following stages he progressed first by improving the means of production and later by developing his powers of reasoning. The current stage was the ninth, beginning with Descartes' philosophy and culminating in the foundation of the French Republic. The tenth and final stage would be government by scientists. Confidence in the future progress of mankind was also expressed by the scientist Joseph Priestley (1733–1804), who sought refuge in America after his house, library, and laboratory had been burnt by a Birmingham mob because of his sympathy for the French revolutionary leaders. There he sought to found a libertarian utopia.

A more profound view of history was developed in 1784 by Immanuel Kant (1724–1804), who in *An Idea for a Universal History from a Cosmopolitan Point of View* argued that, although man wills concord, nature knows better what is good for the species and consequently wills discord. Indeed, the difficulty with the teleological theory of progress that was so widely accepted not only by Condorcet but by many others in the eighteenth and nineteenth centuries was that its validity depended on the unquestioned acceptance of the end to which it was directed, since most progressivists 'unconsciously arrogated to the present a kind of divine right which was arbitrarily denied to all other periods of time'.[12] As Montesquieu and others have pointed out, to carry over into other centuries all the ideas of the period in which one lives is one of the most fertile sources of error.

A pessimistic view of history, and an over-optimistic view of human nature, was taken by Rousseau. Soon after he had rejected modern science and civilization in his prize essay of 1749 he threw away his watch, perhaps influenced subconsciously by the fact that his native city, Geneva, was one of the two main centres (the other being London) of the clock industry. It has been said that no doubt Rousseau would have been happy in Samuel Butler's *Erewhon*, where the mere possession of a watch made one liable to imprisonment.[13] M. J. Temmer has drawn attention to the decisive influence on Rousseau's thought of his reading of St Augustine and hence his love of the static eternity of the Augustinian paradise, and also to Rousseau's 'elegiac desire to live the future in the mode of the past'.[14]

Rousseau's arch-enemy Voltaire was also not a progressivist and even rejected the idea of physical and biological evolution. In an essay of 1769 he argued that the earth has always remained as it was when first created, except for the effects of 'the hundred and fifty days of the Deluge'. As for the marine fossils that had been found on the slopes of Mont Cenis, he attributed their presence so far from the sea to one or other of three alternative possible causes: either collectors had deliberately placed them there, or farmers had brought them there in loads of lime to fertilize the soil, or pilgrims on their way to Rome had accidentally dropped the cockleshell badges from their hats.[15]

The greatest philosopher of history in the eighteenth century was an obscure and poorly paid professor of rhetoric in the University of Naples, Giambattista Vico (1668–1744). Although in his early years he was a follower of Descartes, he gradually came to realize that Cartesianism is applicable as a method only to mathematics and logic and not to our understanding of the external world of nature and society. He not only rejected Descartes's negative attitude to history but he also discarded Descartes's principle of universal doubt as a philosophical starting-point. Vico began instead with the novel idea that, in order to know anything, we must have created it ourselves. Isaiah Berlin has drawn attention to the partial anticipation of this idea by Cardinal Nicholas of Cusa (1401–64), who remarked that mathematics was purely a human creation which we know because we alone have made it, but he did not go on, as Vico did, to extend this insight to historical knowledge and other humane studies.[16] On the basis of what is often referred to as 'Vico's principle of the equivalence of *verum* and *factum*', he argued that, whereas the world of nature can only be fully understood by God, man can understand mathematics because it is his own creation. Similarly—and here he clashed fundamentally with Descartes—Vico believed that history is also understandable, because social institutions, languages, customs, and laws have all been developed by the actions of men, without being pre-ordained. Vico respected mathematics and recognized its value in our attempts to understand the physical world, but he did not identify the two. Moreover, he believed that, since human nature is governed by free will and caprice, mathematical methods cannot be effectively applied to it, or at most only in very circumscribed ways. Vico's originality can be more readily appreciated by us today, although so much of what he advocated now seems commonplace, when we recall that a hundred years after his time Auguste Comte still sought to call his own new science (sociology) by the name of 'social physics', and later still, in 1872, Walter Bagehot gave to his well-known book on the evolution of society the title *Physics and Politics*. (Bagehot seems to have had a peculiar idea of 'physics', since the subtitle of his book is *Thoughts on the application of the principles of 'natural selection' and 'inheritance' to political society*.)

Vico's masterpiece was his book *Scienza nuova* (the title may have been suggested to him by Bacon's *Novum organum* of about a century earlier). The first edition was published in 1725 and the third (revised) edition in 1744. In it Vico maintained that man is a being who can only be understood historically. In other words, knowledge of the past is vital to an understanding of ourselves. He particularly objected to the tendency of reading back into the minds of primitive people modes of thought and feeling that are themselves the product of a long period of historical development. Vico believed that every theory must start from the point where the subject of which it treats began to take shape. As Isaiah Berlin has pointed out, 'This is the whole doctrine of historicism in embryo.'[17] Vico admired the sixteenth-century historian Bodin, who had partially anticipated him by noting that fables and myths can often provide useful evidence of the beliefs of primitive peoples.

Although Vico believed in the existence of historical cycles, he interpreted this concept in a more sophisticated way than had previous believers in it. He thought that certain periods of history had a general basic nature, influencing every detail, which reappeared in certain other periods, so that it was possible to argue by analogy from one such period to another. For example, he drew a parallel between the barbarism of the Christian early Middle Ages in western Europe and the barbarism of the Homeric age, pointing out certain common features, such as rule by a warrior-aristocracy, a ballad-literature, etc. He called such periods 'heroic'. He did not think that history is strictly circular, because novelties are always being created. As R. G. Collingwood has said of Vico's concept of recurrence, 'it is not a circle but a spiral; for history never repeats itself but comes round in each new phase in a form differentiated by what has gone before'.[18] Thus, the barbarism of the western Middle Ages differed from that of Homeric Greece through the influence of Christianity. Vico thought, however, that similar periods tend to recur in the same order; for example, a heroic period is always followed by what he called a 'classical period', in which thought prevails over imagination, prose over poetry, and so forth.

Scienza nuova was very obscurely written and was long neglected. It became famous only about a hundred years after its first publication, when it was discovered almost by accident on a trip to Italy in the 1820s by the great French historian Michelet, who translated it into French and thereby created Vico's reputation. In his monumental *History of France* Michelet declared that Vico did for history what Newton half a century earlier had done for physics. Even if this comparison may seem excessive, there is no doubt that Vico can be regarded as the first exponent of the modern belief that, in order to understand the nature and structure of society, we must study all of its aspects in historical perspective, that is, from the standpoint of time. Although no English translation of *Scienza nuova* appeared until after the Second World War, in the 1920s and 1930s, the Oxford philosopher and historian of Roman Britain R. G. Collingwood was greatly influenced by Vico, but in this he was almost alone in the English-speaking world. Nowadays Vico has come to be regarded as the foremost Italian philosopher and one of the greatest philosophers of history of all time.

A prominent German philosopher of history in the eighteenth century who appreciated the fundamental importance of historical perspective was Johann Gottfried Herder (1744–1803). He rejected 'absolute values' and argued against there being invariant laws of history which are valid for all people and all times. Instead, he believed in 'historical relativism', according to which every culture (and every age) has its own character and intrinsic value by which alone it should be judged. Herder's main work, in four volumes, published from 1784 onwards, was translated into English in 1800 by T. O. Churchill under the title *Outlines of a Philosophy of the History of Man*. Unlike Vico, Herder's influence was soon felt by historians, and this century his ideas were reformulated and elaborated by Oswald Spengler. Isaiah Berlin has provided a stimulating critical account of Herder.[19]

Whereas Vico had restricted the scope of his 'new science' to the history of society and the field of the humanities, the philosopher Immanuel Kant (1724–1804) maintained that *only* the physical universe (excluding biological species which he was convinced were *not* subject to evolution) was the product of continual change and development.[20] Herder, however, was convinced that the historical process embraced the lot, namely, the physical universe, the living world, and human society, although it must be stressed that his point of view 'constituted a philosophical vision rather than a scientific theory'.[21]

Not surprisingly, Herder was extremely critical of the anti-historical prejudices of the French *encyclopédistes*, for example as revealed by their dismissal of Homer as 'a Greek philosopher, theologian, and poet', whose epic poems were 'unlikely to be read much in the future'![22] As the intellectual forerunners of the French Revolution of 1789, the *encyclopédistes* inspired widespread belief in the existence of certain fundamental *timeless* truths which led the more fanatical revolutionaries to proclaim that they were laying down laws not merely for France but for the entire universe!

10. Evolution and the Industrial Revolution

The evolutionary universe

Although Vico laid great stress on the need to view man historically, he did not regard man as emerging out of nature, nor did he think that the natural world had a history of its own. During the course of the eighteenth century, however, the belief began to spread that the idea of time is an essential part of the idea of nature. Just as acceptance of the Copernican theory had shattered the tightly knit confines of the world in space, so similarly the tendency to look at things historically led to a correspondingly vast extension of the world in time.

In his revolt against the then prevailing Aristotelian philosophy of nature, Descartes, like Newton half a century later, regarded all matter, both terrestrial and celestial, as subject to the same physical laws. As a mechanical determinist, however, he did not invoke divine intervention to explain the origin of the solar system. In his *Principia* of 1644 he tried to explain the uniformity of direction of motions in the solar system and their approximation to the plane of the ecliptic by his theory of vortices. He assumed that originally the world was filled with matter distributed as uniformly as possible, and he sketched out qualitatively a theory of successive formation of the sun and planets, including the earth, which he regarded as composed of a series of different layers.

Descartes' idea of the universe evolving by natural processes of separation and combination was the source of a succession of theories of cosmic evolution. Nearly a century later, Swedenborg, in his *Principia* of 1734, advocated a modified view of the Cartesian cosmogony. He suggested that the planets were ejected from the sun, but his idea of how this may have happened was rejected by Buffon who, in 1745, put forward the first tidal theory of the origin of the solar system. Assuming that comets were far more massive than we believe today, Buffon suggested that a comet colliding with the sun may have torn out sufficient material to form the planets.

Neither Swedenborg nor Buffon applied Newtonian ideas to the problems of cosmogony. The first to do so was Immanuel Kant in his *Universal Natural History and Theory of the Heavens*, published in 1755. He assumed that initially all matter was in a gaseous state spread more or less uniformly, except for some primeval regions of higher density which acted as centres of condensation under the action of gravitation. One such centre was the origin of the solar system. Kant thought that eventually through collisions coplanar circular orbits with motions all in the same sense about the sun could arise. He was mistaken in thinking that this phenomenon was automatically possible, because it contradicts the dynamical principle of the conservation of angular momentum and hence Newton's laws of motion. (This dynamical principle, however, was not formulated in full generality until 1775, by Euler). Laplace's nebular hypothesis, put forward in 1796, was free from this defect, and his primeval solar nebula was assumed to rotate initially. The idea of cosmical evolution, as distinct from the old idea of cosmical cycles, was also suggested by the great pioneer of modern observational astronomy, William Herschel. In a paper published in 1814 he claimed that 'the state into which the incessant action of clustering power has brought the Milky Way at present is a kind of chronometer that may be used to measure the time of its past and future existence'.[1]

One of the obstacles that the idea of evolution had to contend with was the widespread inherited conviction that the range of past time was severely limited. Bible-based chronology had already become a severe strait-jacket for scientists studying the nature of fossils. In the seventeenth century both Steno and Hooke realized that fossils were the petrified traces of former living organisms. They were led to develop a dynamical theory of geological change but were confronted with the difficulty of fitting this into the accepted time-scale. The naturalist John Ray was at first inclined to accept the views of Steno and Hooke about fossils. He suggested that, if Steno were right in asserting that mountains had not all existed from the beginning, then perhaps 'the world is a great deal older than is imagined or believed'. Eventually, however, under the influence of his theological beliefs he changed his opinion about fossils in favour of an inorganic origin and reverted to the traditional, and then still widely accepted, non-evolutionary concept of the natural world. Arthur Lovejoy has drawn attention to the following forthright statement, made by Ray in 1703: 'Consult the evidence of experience; elements always the same, species that never vary, seeds and germs prepared in advance for the perpetuation of everything . . . so that we can say, Nothing new under the sun, no species which has not been seen since the beginning.'[2]

During the eighteenth century scientists and others began to discard the Bible-based chronology of nature. In 1721, Montesquieu in his *Lettres persanes* asked 'Is it possible for those who understand nature and have a reasonable idea of God to believe that matter and created things are only 6,000 years old?' Later that century, Diderot thought in millions of years and Kant suggested that the universe may be hundreds of millions of years old. Buffon, when writing his *Époques de la nature*, published in 1778, privately estimated that the first stages of the cooling of the earth would have required at least a million years.[3] In print he was more cautious and estimated the earth's age as being at least 75,000 years. Some of his ideas were condemned by the faculty of theology of the University of Paris.[4]

In 1788 the geologist James Hutton in his *Theory of the Earth* rejected the sudden catastrophic changes that had been previously invoked to explain the stratification of rocks, the deposition of oceans, etc. He realized that the true scientific approach is not to invoke such *ad hoc* hypotheses but to test whether or not the same agents as are operating now could have operated all through the past. In his view, the world has evolved and is still evolving. In one passage he actually likened it to an organism. He concluded that vast periods of time were required for the earth to have reached its present state, and from his study of sedimentary and igneous rocks he came to his frequently quoted conclusion: 'We find no vestige of a beginning—no prospect of an end.'

The idea of using fossils to establish a chronology of the rocks was first suggested in the seventeenth century by Robert Hooke, but was not acted on for over a hundred years. Towards the end of the eighteenth century, William Smith, an English surveyor who collected fossils, realized that each geological stratum could be recognized by the fossils found in it, and that the same succession of strata occurred wherever the rocks concerned were found. He produced in 1815 the first geological map of a whole country. Meanwhile the science of stratigraphical palaeontology was being founded independently in France by Jean-Louis Giraud Soulavie (1752–1813), who was the first to recognize that the stratigraphical ordering of rocks can be regarded as a chronological ordering.

During the nineteenth century the idea of time as linear advancement finally prevailed through the influence of the biological evolutionists, but the climate of thought that made it possible to contemplate the hundreds of millions of years required for the operation of natural selection to account for the present and past species was prepared primarily by the geologists. It was therefore not surprising that Darwin began his life's work as a geologist, as well as a naturalist. Nevertheless, Darwin's demands on the extent of past time came as a great shock to many, as Sir Archibald Geikie explained some forty years after the publication, in 1859, of *The Origin of Species*. Geikie wrote:

Until Darwin took up the question, the necessity for vast periods of time, in order to explain the character of the geological record, was very inadequately realized. Of course, in a general sense the great antiquity of the crust of the earth was everywhere admitted. But no one before his day had perceived how enormous must have been the periods required for the deposition of even some thin continuous groups of strata.[5]

For measurements of geological time, as distinct from guesses, appeal must be made to physics, and here Darwin met what he believed to be one of the gravest objections to his theory. In 1854 the German physicist and physiologist Helmholtz had suggested that the sun maintains its enormous outpouring of radiation by continually shrinking and thereby releasing gravitational energy, which is converted into thermal energy of radiation. He calculated that the current rate of solar radiation could not have been maintained by the sun for more than about twenty million years. This conclusion was supported by the British physicist William Thomson (who became Lord Kelvin in 1892), who thought that at most this estimate could be lengthened to fifty million years.

In confirmation of his view that the hundreds of millions of years demanded by the geologists could not be allowed, Thomson considered the flow of heat through the earth's crust. He argued that this indicated that the earth must be cooling and must therefore have been hotter in the past. He calculated the epoch at which the earth must have been molten and found that this was between twenty million years and an upper limit which he continually reduced from four hundred million to his final estimate, in 1897, of twenty-four million years.

Kelvin was criticized by the geologists but he received public support from the former (and subsequent) prime minister and amateur scientist Lord Salisbury, in his Presidential Address to the British Association at its meeting in Oxford in 1894. Both Kelvin and Huxley were present. Kelvin confined his remarks after the address to a conventional expression of thanks. Huxley's polite and dignified speech of thanks 'veiled an unmistakable and vigorous protest'.[6] The first man who challenged Kelvin on his own ground as a physicist was his former assistant, the mathematician and engineer John Perry (1850–1920). On reading Salisbury's address, he sent a letter to the weekly science journal *Nature*, where it appeared early the next year.[7] He directed attention to Kelvin's simplifying assumption that the earth's thermal conductivity during its cooling was homogeneous, pointing out that, if in fact this conductivity increased towards the centre, Kelvin's estimate of the earth's age would have to be significantly increased. Moreover, if there were some degree of fluidity in the earth's core, thermal conductivity must be supplemented by convection. Perry was attacked arrogantly by the applied mathematician P. G. Tait and in a more moderate tone by Kelvin, who pointed out that, irrespective of his calculations concerning the earth, the sun's heat limited the terrestrial age to a few score million years at most.

While this controversy was raging the concept of evolution was being extended to the history of the earth–moon system. The importance of tidal friction in this context had already been realized in 1754 by Immanuel Kant in the most remarkable of his evolutionary speculations, the short essay that he wrote on the question 'Whether the Earth has Undergone an Alteration of its Axial Rotation'. The frictional resistance of the earth's surface to the tidal currents in the seas and oceans induced primarily by the gravitational action of the moon is very slow in its action, but it is irreversible and over long periods of time could give rise to great changes in the rotation of the earth and the orbit of the moon. Kant's discussion was not quantitatively correct, but it was the first indication that the time of celestial mechanics is not cyclic. Towards the end of the nineteenth century a more thorough and accurate analysis of the dissipative effects of tidal friction on the earth–moon system was made by Charles Darwin's son Sir George Darwin, who tried to fit his results into the time-scale allowed by Helmholtz and Kelvin. He calculated that the minimum time required for the transformation of the moon's orbit from its supposed initial condition to its present form would be fifty to sixty million years. He realized that the actual period was probably a good deal longer. 'Yet I cannot think', he wrote, 'that the applicability of the theory is negatived by the magnitude of the period required.'[8]

The resolution of these difficulties of time-scale for the age of the earth and of the sun was possible only after the discovery of radioactivity at the end of the nineteenth century and the subsequent investigation of nuclear transformations by Rutherford and others early this century. It is now known that there is a sufficient supply of radioactive elements in the crustal rocks to make the net heat loss from the earth extremely small, and Kelvin's estimate for the age of the earth of a few tens of millions of years has been replaced nowadays by about 4,500 million years. Similarly, it is now generally accepted that the sun's heat is maintained by thermonuclear processes in its deep interior that can continue steadily for thousands of millions of years, the age now attributed to the sun being about 4,700 million years.

Radioactivity is an important example of a natural process that is non-cyclic and an indicator of 'time's arrow', i.e. of the unidirectional nature of time. Discovered by Becquerel in 1896, it was explained by Rutherford and Soddy in 1902 in terms of the spontaneous transformation of atoms. It is a purely nuclear phenomenon that is independent of external influences, the rate of 'decay' of a given amount of a radioactive element such as uranium being proportional to the number of atoms of the element present. Consequently, radioactivity not only indicates time's arrow but can also be used as a means of measuring time. Besides those radioactive 'clocks' in the crustal rocks that help us to estimate the age of the earth, another well-known example, discovered more recently, is the carbon-14 clock in organic material that has proved so useful for archaeologists.

In the nineteenth century the unidirectional nature of time in physics was primarily associated with the second law of thermodynamics. This law, originally formulated about 1850 by Rudolf Clausius and William Thomson, was a generalization of the hypothesis that heat cannot of itself pass from a colder to a hotter body. This law determines the direction in which thermodynamic processes occur and expresses the fact that, although energy can never be lost, it may become unavailable for doing mechanical work. Clausius believed that because of this law the universe as a whole is tending towards a state of 'thermal death' in which the temperature and all other physical factors will be everywhere the-same and all natural processes will cease. Although this particular application of the law was disputed, and is now no longer accepted because of recent advances in cosmology, it was for a time a powerful influence undermining long-established belief in the idea of a cyclic non-evolutionary physical universe.

The role of time in modern industrial society

Since the origin of modern industrial society in the eighteenth century time has come to exercise an ever-growing influence on human life generally and even on the way most of us tend to think. For example, consider the concept of 'anachronism'. In antiquity only the Romans seem to have had any idea of it. In ancient Israel the linear concept of history as the fulfilment of a promise made by God involved no such sense; and among the Greeks few writers, apart from Herodotus, showed any awareness of historical *development*. Turning to the Romans, we find that Virgil's characters, unlike Homer's, have a sense of past and future, and that Horace, in the *Art of Poetry*, pointed out that both costume and language change in the course of time. As regards the evolution of language, Horace influenced Chaucer in *Troilus and Criseyde* (c.1386): 'Ye knowe eek that in forme of speche is chaunge / Withine a thousand year.' Thus, as P. Burke has remarked, with reference to this passage, 'The sense of history in one age stimulated the sense of history in another.'⁹ Although the idea of anachronism appears to have influenced some people in the Renaissance period, it only came to be widely appreciated in the course of the eighteenth century. In particular, before the end of that century it led to the introduction of period costume in the theatre.

Perhaps the most striking effect of the growing importance of time on the way people lived was the introduction of an unprecedented country-wide system of organizing transport. The idea of an omnibus service appears to have been first suggested in the middle of the seventeenth century by Pascal, but the first great advance beyond traditional methods did not occur until more than a hundred years later. Indeed in England as late as the reign of George II (1727–60) the customary speed of land travel was no faster than in the first century BC, when it took Julius Caesar, travelling in the comparative comfort of a litter, eight days to cover a distance of 730 statute miles from Rome to Rhodamus. In 1639 Charles I took four days to ride from Berwick to London, a distance of about 300 miles. Owing to the deplorable state of the English roads, which had been greatly neglected since the Roman occupation ended more than a thousand years before, wheeled traffic almost ceased in winter and most people were marooned in their towns and villages for at least half the year. In the seventeenth century some towns near London had a carrier service to and from the capital, but since the roads tended to be appallingly bad, travelling on them in unsprung coaches must have been quite an ordeal even for the hardiest of travellers!

The introduction of tarred roads and the turnpike system in the course of the eighteenth century certainly made for faster going, but the decisive breakthrough came in 1784 when almost within twelve months a unified network of public transport based on strict timekeeping was introduced throughout the length and breadth of England, the mail-coach system. It was founded by John Palmer, MP for Bath. His coach left Bristol at 4 p.m., drove through the night at the standard speed of ten miles an hour, and arrived—strictly on schedule—at the London General Post Office in Lombard Street at 8 a.m. the following morning. Thomas De Quincey, in his well-known essay on 'The English Mail-coach', refers to Palmer as being responsible for 'the conscious presence of a central intellect, that, in the midst of vast distances—of storms, of darkness, of danger—over-ruled all obstacles into one steady co-operation to a national result,' and in a footnote to 'vast distances' he mentions the case 'where two mail-coaches starting at the same minute from points six hundred miles apart, met almost constantly at a particular bridge which bisected the total distance'. He goes on to inform us that it was the mail-coach that distributed over the land 'the heart-shaking news of Trafalgar, of Salamanca, of Vittoria, of Waterloo'. Foreigners often complained of the English mania for saving time. An American Quaker, John Woolman, wrote that 'Stage-coaches frequently go upwards of one hundred miles in twenty-four hours and I have heard Friends say in several places that it is common for horses to be killed with hard driving.'[10]

The introduction of the mail-coach led to a novel problem of time-keeping that was to affect travellers and others for the next 100 years. All towns went by local or 'sun' time, but in the west of England this could be up to twenty minutes behind London's and in the east up to seven minutes ahead. As might be expected, countrymen objected to having London time imposed upon them. The solution that Palmer's Superintendent Hasker adopted was to provide each coach with a timepiece that could be pre-set to lose or gain as required, constant checks being made at certain Post Offices *en route*. The sound of the posthorn was an audible reminder to all inhabitants of the towns and villages through which the mail-coach passed of the importance of time and punctuality. Moreover, the regular sight of the mail-coach must have been a constant reminder to many a countryman of the possibility of seeking his fortune in the town. At the beginning of the nineteenth century four out of every five people in England and Wales were country folk, but by mid-century this was true of only about half.

For most people, travel around the country, either to visit relatives or to go on holiday, had to wait for the advent of railways in the second quarter of the nineteenth century. The effect of steam power on people's way of life and sense of time was not, however, due only to the invention of the locomotive. Steam power was the driving force of the industrial revolution. Although the old cottage-based handloom weavers often had to work very hard for a living, they could at least work when they liked, but factory workers had to work whenever the steam power was on. This compelled people to be punctual, not just to the hour but to the minute. As a result, unlike their ancestors, they tended to become slaves of the clock. The vice of 'wasting time' had already been castigated by Puritan writers, for example by Richard Baxter who, in his *Christian Directory* of 1664, wrote:

> To Redeem Time is to see that we cast none of it away in vain, but use every minute of it as a most precious thing. . . . Consider also how unrecoverable Time is when it's past. Take it now or it's lost for ever. All the men on earth, with all their power, and all their wit, are not able to recall one minute that is gone.[11]

In the nineteenth century this point of view became increasingly widespread, so that even one so remote from manufacturing industry as the poet Wordsworth was criticized by William Hazlitt because he had 'made an attack on a set of gypsies for having done nothing for twenty-four hours'.[12]

Although steam had been used as a source of power for some years, it was not until the Rainhill trials of 1829 with Stephenson's 'Rocket' that it was at last clear that a machine had been produced which was capable of much higher speed than a horse. As Jack Simmons has pointed out, 'the world at large . . . became aware of the railway at a single moment of time'.[13] The same point was emphasized by C. F. Adams, jun. in the 1886 edition of his book on *Railroads*: the locomotive and the railroad 'burst rather than stole or crept upon the world. Its advent was in the highest degree dramatic. It was even more so than the discovery of America.'

At first railways tended to be run in a rather happy-go-lucky manner, timekeeping being the sole responsibility of the engine driver. In 1839, when George Bradshaw was compiling his first railway timetables, one director refused to supply him with the times of arrival of trains, because he believed that 'it would tend to make punctuality a sort of obligation',[14] but the obligation had to be accepted when mail began to be carried. Each town still kept local time, but owing to the greater speed of railway trains than that of the mail-coach the situation became more difficult to control. In Paris clocks outside railway stations were kept five minutes ahead of those inside, not just to ensure that passengers boarded trains in good time but because railway time was Rouen time. In *The Times* of 11 July 1972 there appeared a letter in which the writer said that her late husband, Sir Shane Leslie, had told her that when the famous Provost of Trinity College, Dublin, Professor Mahaffy, once missed a train at a country station in Ireland he observed that the time on the clock outside the station differed from that on the clock inside. When he tackled an elderly porter about this inefficiency, which had caused him to lose his train, the old man scratched his head and replied, 'If they told the same time, there'd be no need to have two clocks!'

In England a uniform railway time was adopted by the middle of the nineteenth century. This was based on Greenwich Mean Time, that is, the time on the meridian of the Royal Observatory at Greenwich, usually denoted by the letters GMT. The Astronomer Royal of the day, Sir George Airy (1801–92), who was the prototype of the modern government scientist, wished to change the attitude of the public to accurate timekeeping. In the late 1840s he was consulted in connection with the design of Big Ben, the huge clock that was to be installed in the tower of the new Palace of Westminster. (It was *not* named after the Chief Lord of Works, Sir Benjamin Hall, but after the prize-fighter Benjamin Caunt, who in his last fight weighed 238 lb. The term 'Big Ben' was often used for an object that was the heaviest of its kind.) Airy insisted that the new clock be regulated by Greenwich time and that the first stroke of the hour should be correct to within one second, an accuracy previously unheard of in a turret clock.

Since the days of Maskelyne all marine chronometers had been tested and checked at the Royal Observatory. In 1833, when John Pond was Astronomer Royal, the Time Ball service was installed there, whereby a ball on the turret of Flamsteed House fell at exactly 1 p.m., so that the time kept by ships on the Thames near Greenwich could be checked thereby. Airy greatly expanded the public service based on GMT by arranging for that time to be distributed throughout the country by means of electric signals. These were transmitted in cables alongside railway tracks, so that for years GMT was called by most people 'railway time'. In his Annual Report of 1853 Airy wrote, 'I cannot but feel satisfaction in thinking that the Royal Observatory is thus quietly contributing to the punctuality of business throughout a large portion of this busy country.'[15]

The advent of railways greatly influenced the family habit of taking an annual holiday, a custom that had previously been restricted to the wealthy. It was the growth of this habit that led to the development of seaside resorts. Not everyone, however, welcomed the new mode of transport and the changes that it produced. For example, when in 1844 the first excursion train to Cambridge was planned, the prospect of an influx on a Sunday of 'foreigners and other undesirable characters to the University of Cambridge on that sacred day' was so unwelcome to the Vice-Chancellor of the time that he wrote to the Directors of the Eastern Counties Railway to complain that 'such a proceeding would be as displeasing to Almighty God as it is to the Vice-Chancellor of the University of Cambridge.'[16]

The revolution in transport affected the tempo of many forms of human activity, particularly the dissemination of news. Although the origin of newspapers, in England at least, can be traced back to the pamphleteering of the different factions at the time of the civil war in the 1640s, it was not until the closing years of the eighteenth century, with the introduction of the mail-coach, and the nineteenth century, with railways, that it became possible for the latest news and informed comments on it to be brought rapidly to towns and villages throughout the land. This spreading of fact and comment far and wide was, of course, also greatly facilitated by the abolition, in the middle of the nineteenth century, of that heavy tax on knowledge, stamp duty on newspapers.

The unprecedented speeding-up of communication, both nationally and internationally, following the introduction of telegraphy and the laying of the transatlantic cable in 1858, revolutionized the conduct of government at home and abroad. An ultimatum could be sent off in the heat of the moment demanding an immediate reply, public opinion could be rapidly influenced and armies mobilized overnight. Such was the march of progress that sudden panic on the New York Stock Exchange in the afternoon could lead to a businessman in London shooting himself before breakfast the following morning. With the advent of wireless telegraphy early in the present century, the rate of dissemination of information all over the world became even more rapid and widespread. No major catastrophe, however remote, now fails to produce agonizing all over the world as soon as it has happened and indeed often while it is still going on.

During the nineteenth century people's attitude to time in countries like England was greatly influenced by the Victorian work-ethic, which led to 'spare time', that is, the time when in principle one was free to do as one liked, being regarded as a reward for hard work. This 'spare time' came to be regulated by the day, week, and year. Previously, holidays had been the forty or more holy days that occurred intermittently throughout the calendar. In England the Puritans, who were in power for over a decade in the middle of the seventeenth century, regarded the traditional Christmas festivities as a pagan relic. They tried to abolish them, but they were soon restored after Charles II returned in 1660. On the other hand, in Scotland Puritan influence persisted and Christmas became far less a time for general celebration than the New Year, a tradition that continued into the present century. The industrial revolution led, however, to the general abolition of holidays based on religious festivals because it was uneconomic to have plant that was expensive to maintain frequently lying idle. In place of the former holy days, four compulsory 'bank holidays' were eventually instituted by law, and it gradually became customary for workers to be given annual holidays of a week or more in the summer. Physical recreation, such as football, came to be organized on a weekly basis, usually on Saturday afternoons.

The nineteenth century saw a great proliferation of pocket watches, although the most important improvement in their mechanism (apart from the balance spring) had already been introduced the previous century. This was the lever escapement invented by Thomas Mudge (1715–94). Later, the mechanism of watches was further improved by Abraham Louis Breguet (1747–1823), who also designed, in 1815, an observatory clock to strike each second—the forerunner of the modern time-signal. A prominent early nineteenth-century English horologist who had an important and lasting influence on watchmaking in other countries, notably France and Switzerland, was John Arnold (see p. 146). By the middle of that century Sir John Bennett, whose firm had been founded in 1843, recognized the danger of the growing competition from the Swiss watchmaking industry. He therefore arranged for watch mechanisms to be imported into England so that his firm could put the necessary finishing touches to them and sell them as British. He spent lavishly on advertising his wares at the Great Exhibition of 1851. Later in the nineteenth century the modern mass-production of watches began in USA, but it was taken up and greatly extended by the Swiss, who soon dominated the industry.

One of the most surprising facts in the history of horology is that, long after the invention of more precise devices, makers of domestic clocks and watches continued to make use of the verge escapement. This was because it proved to be particularly well-suited for withstanding the rigours of domestic use and portability, whereas escapements such as the anchor type needed to be kept on level surfaces if they were to function satisfactorily.

Not only do most workers nowadays have to clock in and clock out when they begin and end their working day, but timekeeping applies no less generally to sporting activities. Indeed, anything, however idiotic, can now be regarded as a sport so long as it can be timed and can be used to set up a 'record'. Kevin Sheenan, of Limerick, acquired a kind of fame by talking non-stop for 127 hours, and in the USA a preacher established another record by delivering a sermon that lasted forty-eight hours. (This achievement would not have amused Queen Victoria who is said to have had placed conspicuously in all the pulpits used by her chaplains a sand-clock that ran for only ten minutes!) In these and many other ways most of us have become more and more subservient to the tyranny of time. As Lewis Mumford has so pertinently remarked, 'The clock, not the steam-engine, is the key-machine of the modern industrial age.'[17] The popularization of timekeeping that followed the mass production of cheap watches in the nineteenth century accentuated the tendency for even the most basic functions of living to be regulated chronometrically: 'One ate, not upon feeling hungry, but when prompted by the clock; one slept, not when one was tired, but when the clock sanctioned it.'[18] A good example of how strange our modern preoccupation with time seemed to someone used to a very different way of life is provided by the diary kept by the Nepalese ruler Jang Bahadur on his visit to Britain in 1850. According to the translation by John Whelpton of a biography of him in Nepali published in Katmandu in 1957 and containing excerpts from this diary, he remarked: 'Getting dressed, eating, keeping appointments, sleeping, getting up—everything is determined by the clock . . . where you look, there you see a clock.'[19]

Although by 1855 about 98 per cent of the public clocks in Britain were set to GMT, acceptance of this time generally throughout the country encountered difficulties. For example, in the case of *Curtis* v. *March* at Dorchester assizes on 25 November 1858, the judge took his seat on the bench at 10 a.m. by the clock in the Court, but as neither the defendant nor his lawyer were present he found for the plaintiff. The defendant's counsel then entered the Court and claimed to have the case tried on the ground that it had been disposed of before ten o'clock by the town clock, whereas the clock in the Court was regulated by Greenwich time, which was some minutes before the time in Dorchester. On appeal, the assize judge's decision was reversed, on the ground that 'ten o'clock is ten o'clock according to the time of the place'. This decision was held to define legal time in Great Britain until 1880.[20] In that year *The Times* published a letter from a 'Clerk to the Justices' pointing out the difficulties of officials conducting parliamentary elections in deciding the correct time to open and close the poll. Later that year an Act of Parliament was passed giving legal sanction throughout Great Britain to Greenwich Mean Time.

Soon afterwards steps were taken to standardize timekeeping throughout the world. In 1882 the United States passed an Act of Congress authorizing the President to call an international conference to decide on a common prime meridian for time and longitude, and in October 1884 delegates from twenty-five countries assembled in Washington for the International Meridian Conference. With only one country (San Domingo) voting against and two others (France and Brazil) abstaining, it was agreed to recommend that the Prime Meridian of the world should pass through the centre of the instrument at the Observatory at Greenwich known as the Airy Transit Circle and that the Universal Time should be GMT. This was not surprising, since the invention of the marine chronometer by John Harrison and the introduction of the *Nautical Almanac* in 1766 by the Astronomer Royal Nevil Maskelyne had already led many mariners of all nations to use Greenwich time and the Greenwich meridian. By the early 1880s nearly three-quarters of the ships throughout the world used charts based on the Greenwich meridian. Until 1925, however, as mentioned on p. 15, astronomers continued to begin their day at noon, because it meant that the date did not change in the middle of a night's observing. Another important consequence, although not specifically recommended by the Conference, was the setting-up of a time-zone system throughout the world as had originally been suggested by an American professor, Charles Dowd, in a pamphlet published in 1870. The need to co-ordinate timekeeping was much greater in a large country such as the United States than in Great Britain, but the crucial factor that influenced Dowd was the different times kept by the many railway companies that sprang up after the Civil War and the great inconvenience that they caused to the travelling public. For example, at Pittsburgh, Pennsylvania, there were six different time-standards for the arrival and departure of trains. Dowd's proposal was a scheme identical in principle with the standard time system used throughout the world today.

As long ago as 1881 an American, G. Beard, wrote a book called *American Nervousness* to point out that the widespread and increasing emphasis on punctuality was causing men to worry that 'a delay of a few moments might destroy the hopes of a lifetime'. Among those in Europe who were anxious for time to be standardized was Count Helmuth von Moltke, who pleaded with the German Reichstag in 1891 for the abolition of the five different time-zones in Germany because they severely impeded the co-ordination of military planning.[21] The resulting adoption of a single standard time greatly facilitated German mobilization in 1914. On the other hand, in France, where the lack of time-standardization was much worse than in Germany, a journalist, L. Houllevigue, writing in *La Revue de Paris* in July 1913, admitted that the delay in correcting this until 1911 was primarily due to Anglophobia. Indeed, the law that came into force defined legal time in France as nine minutes and twenty seconds later than Mean Paris Time. 'By a pardonable reticence, the law abstained from saying that the time so defined is that of Greenwich, and our self-respect can pretend that we have adopted the time of Argentan, which happens to be almost exactly on the same meridian as the English observatory.'

One of the main reasons for the catastrophic failure of diplomacy to prevent the outbreak of the First World War in August 1914 was the inability of diplomats to cope with the enormous volume and unprecedented speed of telegraphic communication in the last days of July. The rate at which messages could be sent from one capital to another necessitated rapid and often ill-considered responses. Ironically, the main reason for the failure of the Schlieffen plan for attacking France through Belgium was the unprecedented success of German mobilization, with thousands of trains ferrying troops to the front so rapidly that they outran their own timetable and consequently the supplies they needed failed to keep pace with them.

The decisive weapon in that war was the machine-gun with its rapid firing. On the Western front it has been estimated that it caused four-fifths of the casualties. Of the 60,000 casualties that the British army suffered on the first day of the battle of the Somme, 1 July 1916, most occurred in the first hour—probably in the first few minutes. One of the social consequences of the First World War was the increased use of wrist-watches. Many men had considered them unmanly until they became standard military equipment. The battle of the Somme began when hundreds of platoon leaders blew their whistles as soon as their synchronized wrist-watches showed that it was 7.30 a.m. Thus, whereas Einstein had shown ten years before that in the physical world simultaneity was a 'private' concept rather than a 'public' one (see ch. 11), in the world of human action it had become far more important than it had ever been.

The introduction of the radio time-signal early this century for the dissemination of time for navigational purposes led to the final abandonment of the lunar-distance method of determining longitude at sea, since it now became possible to check a ship's chronometers directly. (The lunar-distance method had been used occasionally to check chronometers at sea when no other method was available.) Since the First World War radio, and later television, and the ever-increasing speed of the new modes of transport made possible by the invention of the internal combustion engine have led to our dependence on the clock becoming ever greater. In recent years the most spectacular example of this has been provided by space vehicles and the associated requirement of ultra-precise timekeeping.

In the early 1920s the accuracy of civil timekeeping was significantly improved by W. H. Shortt, a railway engineer, who in association with the horologist F. Hope-Jones and the Synchronome Company perfected what came to be known as the Shortt free-pendulum clock. The material used for the pendulum was a virtually temperature-independent alloy of steel and nickel called 'invar', first produced some years before in France. Any interference with the free motion of the pendulum was reduced to a minimum by the ingenious use of a subsidiary 'slave clock'. Shortt clocks were the standard timekeepers at the Royal Observatory, Greenwich, from 1925 to 1942. Previously, the best clock-accuracy was about one-tenth of a second (100 milliseconds) a day, but Shortt clocks were accurate to about 10 seconds a year, that is, about 30 milliseconds a day. In the 1930s still greater accuracy was obtained by utilizing the mechanical vibrations of the crystalline mineral quartz, instead of the vibrations of a pendulum in the earth's gravitational field. The quartz crystal clocks which replaced the Shortt clocks as the standard timekeepers at the Royal Observatory in 1942 were accurate to about two milliseconds a day.

For centuries the time kept by our clocks and watches was controlled by the rate of rotation of our planet, but with the invention of more accurate clocks it was found that the rotating earth is not a sufficiently accurate timekeeper for modern needs, because it is subject to small variations. The earth is a solid body surrounded by water and air, and seasonal changes in these, for example the melting and freezing of the polar ice-caps, affect the earth's rate of rotation so that the length of the day fluctuates during the year by just over a millisecond (thousandth of a second). There are also small irregular changes attributed to processes in the earth's deep interior. Besides these changes there is a progressive slowing down of the earth's rate of rotation caused by tidal friction in shallow seas that produces an increase in the length of the day of about 1.5 milliseconds a century. As a result in 1952 the rotating earth was displaced as the fundamental timekeeper by Ephemeris Time, based on the length of the year, which is decreasing by about 0.5 seconds a century but can be predicted. However, even this did not prove entirely satisfactory and because of the increasing demand for high-precision time-measurement it has become essential to have some more fundamental standard of time than any that can be derived from astronomical observations. Such a standard is given by the frequency of a particular spectral line of an atomic or molecular vibration. The most successful method of this type has been developed by Dr L. Essen of the National Physical Laboratory.[22] Consequently, in 1967 a new definition of the second was made in terms of the electromagnetic radiation generated by a particular transition in the ground state of the caesium atom. It is called the 'SI second' (Système Internationale). It is formally defined as the duration of 9,192,631,770 periods of radiation corresponding to the transition between two hyperfine levels of the caesium-133 atom. In this transition the spin of the outermost electron of the atom 'flips over' with respect to the spin of the nucleus. (A quartz crystal oscillator is controlled by means of a known relationship between its frequency and that of the radiation generated by this transition.) The caesium atom was chosen because the frequencies concerned are in the radio range and can be measured by standard techniques. In recent years there has been much technical discussion concerning the relation between TAI ('International Atomic Time'), obtained by the continuous summation of time-intervals deduced from these measures of frequency, and the time-scales used by astronomers. The accuracy of astronomical time, which is still required for practical purposes, is checked by means of the frequency of this radiation. This atomic frequency standard is now so precisely determined that in individual cases its accuracy can be as high as one part in 10^{14}, which is equivalent to an error of only one second in three million years.

The world's time signals are now co-ordinated by the Bureau International de l'Heure (BIH) based on a world 'mean clock' that is the average of some eighty atomic clocks in twenty-four countries. It provides direct synchronization to within about a millisecond. Although this 'Co-ordinated Universal Time' (UTC), which has replaced GMT as the basis of civil time throughout the world, is now controlled from Paris, the world's prime meridian for longitude and time still passes through the old Observatory at Greenwich. In practice, the zero meridian is now defined by the adopted longitudes of the instruments that contribute to the determination of UTC. Since 1985 the contribution of the Royal Greenwich Observatory to the international determination of UTC and longitude has been through its observations of the artificial satellite *Lageos* by a laser-ranging system in use at Herstmonceux since the autumn of 1983. As from 1 January 1972 time signals have radiated atomic seconds, but just as there is not a whole number of days in a year, so there is not a whole number of atomic seconds in a solar day. This has led to the adoption of corrections, either positive or negative, of exactly one second. They are called 'leap seconds' and when required are on the last day of a calendar month, preferably on 31 December or 30 June.

11. Rival Concepts of Time

Instant and duration

St Augustine appears to have been the first thinker to have carefully investigated the consequences of our actual experience of time being confined to the present instant. He came to the conclusion that our ideas of past and future must depend on our consciousness of memory and sense of expectation. In regarding time from this psychological point of view the primary concept is the instant rather than duration. Nevertheless, despite St Augustine's great influence on medieval theology, it was not until the humanistic Renaissance of the fifteenth century and the religious Reformation of the sixteenth century, followed by the Copernican revolution in astronomy and cosmology—all of which contributed to the dissolution of the timeless medieval world-picture with its hierarchical structure in which everything had its assigned place—that Western thinkers began to regard personal existence as being essentially based on the present moment.

The significance of the instant was presented in pictorial art by Hans Holbein (1497–1543), for example in his famous painting of 1533 known as 'The Ambassadors'. In a pamphlet on this painting, published by the Trustees of the National Gallery in 1974, Alistair Smith has emphasized the sense of *instaneity* at the centre of Holbein's art. Fascinated by the nature of human mortality, his object was to depict the sense of personal existence at a definite instant. Smith draws attention to the way in which this moment of time was precisely indicated in the picture, the date (11 April) being registered on a cylindrical dial and the hour (10.30 a.m.) on a polyhedral dial.

One of the first to give literary expression to the notion of personal existence as based on the present moment was the famous French essayist Michel de Montaigne (1533–92). Even as a child he is said to have been greatly influenced by the *Metamorphoses* of Ovid, and his outlook throughout life was dominated by the conviction that the world in which we find ourselves is in a state of incessant change. Consequently, he believed that the assumptions on which our way of thinking is based are necessarily uncertain and defective.

This attitude of scepticism concerning human knowledge was turned to positive effect later by René Descartes (1596–1650), in the philosophy that he developed on the basis of his famous axiom *Cogito ergo sum*. If, however, existence is to be identified with the transitory instant rather than with duration, how can the continued existence of the world be accounted for? Descartes's answer was that the world is recreated from instant to instant, conservation and creation differing only in respect of our way of thinking and not in reality, self-conservation being the unique prerogative of God.

In the eighteenth century, however, there was a general revolt against the idea of the instant as the basic temporal concept. Instead, it came to be appreciated that our experience of time is dualistic: intensity of sensation is associated with the instant, but our awareness of multiplicity of sensation depends on duration. This led to a new interest in the nature and significance of memory. For example, the French *philosopher* Denis Diderot (1713–84) in a famous passage (*Oeuvres*, IX, p. 366) wrote:

I am led to believe that everything we have seen, known, perceived, heard—even the trees of a deep forest—nay, even the disposition of the branches, the form of the leaves and the variety of the colours, the green tints and the light; the look of grains of sand at the edge of the sea, the unevenness of the crests of waves, whether agitated by a light breeze, or churned to foam by a storm; the multitude of human voices, of animal cries, and physical sounds, the melody and harmony of all songs, of all pieces of music, of all the concerts we have listened to, *all of it, unknown to us, exists within us.*[1]

This remarkable claim, for which Diderot had no scientific evidence, has been supported this century by the experiments made by the Canadian neurosurgeon Wilder Penfield, who elicited 'flashbacks' by applying an electrode to the exposed cortex of patients undergoing brain surgery.[2] (See *The Natural Philosophy of Time*, pp. 103 ff.)

In the nineteenth century the idea of temporal succession came to assume greater importance in human life and thought than ever before. It not only gave rise to important developments in literature, including the evolution of the novel and a spate of autobiographies, but it also became of dominating importance in the natural sciences, as is exemplified by the oft-quoted statement of the geologist G. J. P. Scrope: 'The leading idea which is present in all our researches, and which accompanies every fresh observation, the sound of which to the ear of the student of Nature seems continually echoed from every part of her works, is—Time!—Time!—Time!'[3]

As the century progressed, we find that truth itself tended to be regarded no longer as eternal and changeless but as time-dependent. Attention came to be focused on the historical process rather than on an eternally valid, unchanging order of things. In other words, interest was transferred from the 'thing completed' to the genetic process, that is, from 'being' to 'becoming'. This radically new point of view received its extreme formulation in the philosophy of the 'modern Heraclitus', Henri Bergson (1859–1942), for whom ultimate reality was neither 'being' nor even 'being changed' but the continual process of 'change' itself, which he called *la durée*. An authoritative critical account of Bergson's eloquently expressed philosophy of *la durée* and its influence in the early decades of the present century has been given by the distinguished former Professor of the History of Philosophy in the University of Warsaw, who is now a Fellow of All Souls, Leszek Kolakowski in his book *Bergson*, published in 1985 by Oxford University Press, in the 'Past Masters' series. Bergson achieved the unique distinction of being both scathingly criticized by Bertrand Russell (in 1912) and having his books placed on the *Index Prohibitorum* by the Holy Office in 1914—the year in which he was elected a member of the Académie Française! A more scientifically orientated philosophy of change than Bergson's, but which owed something to his example, was developed between the wars by the British mathematician and philosopher A. N. Whitehead (1861–1947), particularly in his book *Process and Reality*, based on his Gifford Lectures of 1928.

Relativistic and cosmic time

In view of the important role that time had come to play in modern life as well as in the scientific world-view, it was a great surprise when, in 1905, in a scientific paper that is now regarded as one of the most important published this century Albert Einstein revealed a previously unsuspected limitation of the current theory of time. According to that theory, for a given way of measuring time each event can have only one time associated with it. Events having the same time are said to be 'simultaneous'. The new point that occurred to Einstein was that, although the idea of simultaneity is perfectly clear for two events occurring at the same place as well as at the same time, it is not equally clear for two events occurring at different places. Instead, the simultaneity of a distant event and one occurring in the observer's own experience depends on the relative position of the distant event and the mode of connection between it and the observer's perception of it. If the distance of the event is known and also the velocity of the signal (e.g. light) that connects it and the observer's perception of it, the observer can correlate the event with some previous instant in his own experience and can regard these two events as simultaneous. This calculation will, of course, be a separate operation for each observer, but until Einstein raised the question it had been tacitly assumed that, when such calculations are correctly performed, all observers will agree on the time of any given event. Einstein produced a successful theory in which this is not the case.

Einstein based his special theory of relativity, as it came to be called, on the principle that the laws of nature are expressible in the same mathematical form for all observers in uniform relative motion (including relative rest). This principle of relativity holds good in classical dynamics based on Newton's laws of motion, but Einstein believed that it should be extended to other branches of physics, in particular electromagnetism and the theory of light. In classical dynamics there is no velocity with special properties, but in electromagnetic theory the velocity of light (in empty space), which is about 300,000 kilometres a second, has special significance. Einstein believed that, if the properties of light are to be the same for all observers in uniform relative motion, they must all assign the same velocity to it. This additional condition, however, he found to be incompatible with the prevailing theory of time. Although according to his theory any two observers at relative rest will assign the same time to any given event, wherever it may occur, this is found not to be the case for any two observers in uniform relative motion in general. Consequently, the condition that each event has only one time associated with it no longer holds. Instead, its time depends on the observer.

Einstein's theory involves the assumption that no physical effect can be transmitted faster than the velocity of light (in empty space). Although neither Newton nor Leibniz imposed any such restriction, Einstein's theory is more in accord with Leibniz's concept of time than with Newton's. For, although Leibniz's idea that time is derived from events is compatible with Einstein's theory, Newton's concept of absolute time is not. Whereas for Newton time was independent of the universe and for Leibniz it was an aspect of the universe, the view that now prevails since Einstein's theory has come to be regarded as an essential part of physics is that time is an aspect of the universe which depends on the observer.

An important consequence of Einstein's special theory of relativity is that a moving clock will appear to run slow compared with a similar clock at rest with respect to the observer, and the closer the velocity of the moving clock is to that of light the slower it will appear to run. This apparent slowing down of a moving clock is called 'time dilatation'. Of all the consequences of Einstein's theory this is the one that many people have found it most difficult to accept, since it clashes with our common-sense intuition of time. Nevertheless, there is now abundant experimental evidence, particularly that provided by high-speed particles, supporting this conclusion.

In his 1905 paper Einstein restricted the principle of relativity to observers in uniform relative motion and did not consider gravitational effects. In what he called the general theory of relativity, which he developed some ten years later in order to cope with gravitation, he extended the principle of relativity to include observers in any form of accelerated motion, special relativity being regarded as an important particular form of this more comprehensive theory. In this theory too the classical assumption that each event occurs at a unique time, the same for all observers, does not apply. In view of this, the idea of a unique cosmic time-scale for the physical universe as a whole might be thought to have no objective significance. Such a conclusion would, however, be mistaken, as has been shown by the developments that have taken place this century in cosmology.

In 1924 the astronomer E. P. Hubble, using the then recently installed 100-in. diameter telescope on Mount Wilson in California showed that the general background of the universe is formed not by the stars but by the galaxies, of which the Milky Way stellar system (that includes our own sun) is one. Five years later he found that the galaxies are receding systematically from one another. (The recessional motion of a galaxy is measured by its 'red shift', that is, the displacement of identifiable lines in its spectrum towards the red.) This discovery made almost as great a change in man's conception of the universe as the Copernican revolution four centuries earlier. Instead of an overall static model of the cosmos, it appeared that the universe is expanding, the rate of relative recession of the galaxies being proportional to their respective distances from one another. This is known as 'Hubble's law'.

Hubble's discovery stimulated much work in theoretical cosmology largely based on Einstein's general theory of relativity. As a result, there was a revival of the idea that there are successive states of the universe associated with a world-wide scale of time. This came about because in

each of the world-models considered there was a definite set of particularly significant hypothetical observers, namely those located on the
individual galaxies and moving with them. The local times associated
with these observers fitted together to produce a world-wide time called
'cosmic time'.

An important assumption in the construction of the expanding world-
models studied after Hubble's discovery of cosmical recession was that
hypothetical observers on each galaxy would see themselves at a centre of
isotropy (i.e. spherical symmetry) of the whole universe, so that its
general appearance in each direction would be the same. Observational
evidence in favour of this assumption can therefore be regarded as
supporting the concept of cosmic time. Impressive confirmation of the
assumption of cosmic isotropy has come from the discovery of what is
often called the 'primeval fireball'. In 1965 A. A. Penzias and R. W.
Wilson at the Bell Telephone Laboratory in New Jersey found that some
unexpected radiation was leaking into the antenna of their radio-tele-
scope. They soon discovered that this radiation was practically isotropic
and at the wavelength at which they were working its intensity was
equivalent to a temperature of about three degrees on the kelvin, or
absolute, scale. This radiation has been interpreted as the relic of the
primeval high temperature radiation associated with an explosive origin
of the universe, a conclusion which has been accepted by most astronomers. The fact that the radiation is highly isotropic rules out the possibility of any local origin of its source. A source restricted either to the
solar system, or our galaxy, or even to the local cluster of galaxies, could
not produce radiation that would appear to us as isotropic. Moreover,
large-scale departures from isotropy anywhere in the universe would
affect the radiation and make it seem anisotropic to us. Consequently,
the isotropy of the cosmic background radiation is powerful evidence
that the universe is basically isotropic about each galaxy, and this is a
strong argument for the existence of cosmic time.

The discovery of the expansion of the universe and the evidence for the
existence of cosmic time have not only reinforced the tendency in recent
centuries for time to become a major feature of the scientific world-view,
but have thrown new light on the old problem of the total extent of past
time. Although in the eighteenth and nineteenth centuries it was becoming clear to those who had discarded an out-of-date biblical chronology
that the age of the universe must be reckoned in hundreds, and possibly
thousands, of millions of years, it was not until the present century that
more precise estimates could be made. As already mentioned (p. 157), the

discovery of radioactivity and the development of nuclear physics have led us to assign ages to the earth and sun of nearly 5,000 million years. Moreover, astrophysicists have reason to believe that the ages of the oldest stellar clusters and our galaxy are between 10,000 and 16,000 million years. As for the universe as a whole, Hubble's law has been used to estimate its age. Observational evidence of the current rate of recession of the galaxies when applied to the simplest expanding world-models indicates that the universe may have had an explosive origin between 10,000 and 20,000 million years ago, the latter probably being the more correct estimate. Despite the uncertainties involved in obtaining this result, it is remarkable that it is consistent with the totally independent estimates of the ages of the oldest star clusters and our galaxy. In the present state of knowledge, it would seem to be the longest stretch of past time over which we can extend the existence of the physical universe as we know it.

12. Time, History, and Progress

Time and belief in progress

Although time has come to play an increasingly important role in modern thought, there have been considerable differences of view concerning 'progress'. The period from about 1750 to 1900 was the age of greatest faith in that concept as well as being an age when people became more and more aware of the significance of time. For example, in Paris in the 1820s the historian Guizot drew vast audiences to his masterly lectures on the history of Europe in which he argued that the fundamental idea embedded in the word 'civilization' is progress. This belief was greatly encouraged by the spread of democracy. As Alexis de Tocqueville pointed out in a famous passage of his classic *Democracy in America*, published in 1835, whereas aristocratic nations are naturally inclined to narrow the scope of human perfectibility, democratic nations tend to suffer from the opposite tendency.

It can scarcely be believed how many facts flow from the philosophical theory of the indefinite perfectibility of man, or how strong an influence it exercises even on men who, living entirely for the purposes of action and not of thought, seem to conform their actions to it, without knowing anything about it. I accost an American sailor and inquire why his ships are built to last for a short time; he answers without hesitation that the art of navigation is making such rapid progress that the finest vessel would become almost useless if it lasted beyond a few years. In these words, which fell accidentally and on a particular subject from an uninstructed man, I recognize the general and systematic idea upon which a great people direct all their concern.[1]

Belief in progress was strongly reinforced by Darwin's theory of biological evolution as presented in *The Origin of Species*, which appeared in 1859. It was also of importance to the other discoverer of the principle of natural selection, Alfred Russell Wallace, and even, more to the engineer, philosopher, and sociologist Herbert Spencer, who tried to make the principle of progress the supreme law of the universe. The inevitability of progress was also an article of faith for Comte, Marx, and other nineteenth-century philosophers of history. In their different ways both Comte and Marx believed in the existence of three successive stages of social evolution; in Comte's case, the theological, the metaphysical, and the 'positivistic' (scientific); and in Marx's case the Hegelian sequence of thesis, antithesis, and synthesis. More than a hundred years before Comte and Marx, Vico also had a concept of three distinct historical stages dominated, respectively, by gods, heroes, and men, in that order. Ruskin even divided geological history into three periods: first, that in which the earth crystallized; then that in which it was 'sculptured'; and finally that now present in which it is being 'desculptured' or deformed, mountains being eroded, glaciers piling up debris, and so on. This historical type of triadic numerology, which has manifested itself in other ways too—for example, Moscow as the 'Third Rome' and Hitler's Germany as the 'Third Reich'—can be traced back to Joachim of Fiore in the thirteenth century.

   Nineteenth-century belief in the reality of progress was accompanied by an increased awareness of the importance of history for the under-standing of subjects such as law. The historical attitude to jurisprudence can be traced back to Gustav von Hugo (1764–1844) and Friedrich von Savigny (1779–1861) in Germany. Savigny argued powerfully against the idea that had been prevalent the previous century, particularly in France, that law can be arbitrarily imposed on a country irrespective of both its current state and its history. In England, Savigny's point of view inspired Sir Henry Maine (1822–88), Professor of Civil Law in the University of Cambridge from 1847. His well-known book *Ancient Law*, first published in 1861, introduced into this country the idea of the historical approach not only to the study of law but also to that of society generally. The outstanding figure in developing the historical approach to the study of society was, however, the Oxford anthropologist E.B. Tylor (1832–1917), the first edition of whose famous book *Primitive Culture* appeared in 1871. Although he explicitly drew attention to the persistence of pre-scientific modes of thought in civilized societies, he argued that the history of man, as revealed by a study of the implements he has used, is indubitably 'the history of an upward development'. He maintained that the essence of progress is man's *intellectual* development, since it is the precondition of his progress in all other respects. Despite their awareness of the many difficulties and setbacks that had occurred in the history of mankind, belief in the reality of progress and consequently in the beneficent nature of time was uppermost in the minds of many prominent Victorian thinkers.

Nevertheless, whereas a poet such as Tennyson in 'Locksley Hall' (1842) could write,

> Not in vain the distance beacons. Forward, forward let us range,
> Let the great world spin for ever down the ringing grooves of change,
> Through the shadow of the globe we sweep into the younger day;
> Better fifty years of Europe than a cycle of Cathay.

there were other writers who came to feel the menace of time as much as its promise, and poets such as Blake and Shelley and, more recently, Yeats persisted in the belief that there are cycles of civilization. Similarly, the philosopher Nietzsche, who died in 1900, and the twentieth-century historians and sociologists Spengler, Pareto, and Toynbee all believed in the cyclical nature of history. Meanwhile, criticism of the superficial optimism that had been encouraged by the widespread popularization of Darwinism began to be voiced, for example by the philosophically minded statesman A. J. Balfour in his Rectorial Address to the University of Glasgow in 1891, in which he pointed out that 'the theory of evolution does nothing to justify optimism about the future of mankind'. A similar view was expressed by 'Darwin's bulldog', T. H. Huxley. A generation later Dean Inge in his Romanes Lecture 'The idea of progress', delivered at Oxford in 1920, made the caustic comment: 'The European talks of progress because by the aid of a few scientific discoveries he has established a society which has mistaken comfort for civilization.' Loss of faith in progress had already been made the subject of an acute historical analysis, in 1908, by the French writer Georges Sorel in his book *Les illusions du progrès*.

Following the First World War the general climate of opinion, particularly in Germany, caused the pessimistic views of Oswald Spengler to attract considerable attention. His widely read book *The Decline of the West* seemed to many British readers more cogent than the old-fashioned optimism expressed by J. B. Bury in his book *The Idea of Progress*, published in 1920. Spengler's philosophy of history was based on the extension of Goethe's morphological concept of organic nature to include 'cultures'. Spengler regarded these as being 'plant-like', that is, subject to generation, growth, and decay and confined to particular regions of space. Unlike his British counterpart Arnold Toynbee,

Spengler paid particular attention to the influence of science and technology on history, although he interpreted this influence in his own peculiar cyclical way.

In an important collection of essays on the social and philosophical
impact of modern science published in 1978 with the title *Paradoxes of
Progress*, the molecular and neurological biologist Gunther Stent, of the
University of California (Berkeley), has carefully analysed the meaning
of 'progress'. He rejects the traditional interpretation of this concept in
terms of 'greater happiness' and the 'perfectibility of man', because no
precise meaning can be attached to these ideas. Instead, he argues that the
only useful way to analyse 'progress' is in terms of the 'will to power'. In
other words, a 'better' world signifies 'one in which man has a greater
power over external events, one in which he is economically more
secure'.[2] In Stent's view only this definition can make progress 'an
undeniable historical fact'. Stent believes, however, like Spengler, that
scientific progress may be coming to an end. A better grounding in the
history of science might have saved him from coming to this pessimistic
conclusion, for at the end of the eighteenth century the great mathematician Lagrange thought that mathematical discovery was on the point of
petering out, although in fact the following century and the present have
seen a greater proliferation of mathematics than occurred in all previous
centuries. Similarly, towards the end of the nineteenth century many
physicists believed that little new remained to be done in their subject,
except to clear up a few minor anomalies and obtain still more accurate
values for the principal physical constants. In a lecture delivered at the
Royal Institution in 1900 with the title 'Nineteenth century clouds over
the theory of heat and light',[3] however, Lord Kelvin considered two to
be important. His concern was not misplaced, particularly as it eventually turned out that one of these clouds would require for its dispersal
the theory of relativity developed by Einstein in his famous paper five
years later. Moreover, in the same year that Kelvin delivered his lecture
Planck introduced the quantum hypothesis. These fundamental
advances, together with Rutherford's contemporaneous experimental
researches on radioactivity, heralded the golden age of modern science
characterized by an unprecedented increase in knowledge and in man's
command over nature. In the context of this historical perspective we are
justified in querying Stent's conclusion that scientific progress may be
coming to an end. Moreover, irrespective of whether scientific and
technological discoveries are used beneficially or otherwise, the knowledge that results from them is cumulative—and will remain so unless all

civilization on this planet comes to a catastrophic and complete end. Consequently, even if we can no longer put our trust in the simple faith of Joseph Priestley, the discoverer of oxygen, who saw in the history of science an exemplification of what, in the preface to his *The History of Electricity with Original Experiments* (1767) he called a 'perpetual progress and improvement' that would continue to heights that are 'really boundless and sublime', there is now no doubt that the continuing momentum of scientific, medical and technological progress makes it impossible for our civilization to be regarded as either static or cyclic.

Time, history, and the computerized society

There is good reason for believing that we are now in the early stages of one of the main irreversible changes in the history of man as we enter the computer age. No longer will it be appropriate to regard the clock as the only key-machine of the modern industrial age. In the Preface to his illuminating book *Turing's man: Western Culture in the Computer Age* (1984) David Bolter writes that 'It makes sense to examine Plato and pottery *together* in order to understand the Greek world, Descartes and the mechanical clock *together* in order to understand Europe in the seventeenth and eighteenth centuries. In the same way, it makes sense to regard the computer as a technological paradigm for the science, the philosophy, even the art of the coming generation.'[4] The computer thus joins the clock as one of the *two* key-machines of the coming new technological age. In the current use of the word, the 'computer' is no longer just a machine for effecting numerical calculations but is far wider in its scope: in short, it has become a machine for processing all kinds of information. As Bolter points out (p. 109), 'The computer programmer is concerned about time because he wants to get a job done . . . All the elaborate mathematization of time comes down to the desire to put time to work.' Once a new technology has been invented, it tends to proceed with its own relentless logic and thus may have a lasting effect on a whole civilization. We have seen that this is what happened after the invention of the mechanical clock and this is what is now happening since the deep mathematical insight of Alan Turing (1912–1954) and of J. von Neumann (1903–1957) has led to the invention of the modern digital computer, perhaps the greatest achievement of twentieth-century technology.

The general concept of the modern computer was due originally to Charles Babbage (1792–1871) and Byron's daughter Lady Lovelace, but a century later Turing and von Neumann realized that it would be possible to succeed with the aid of electronic components, including the transistor, which was invented in 1949, where Babbage with mechanical gears had failed. One of the great differences is the fantastic speed at which the modern computer works, the times now used to release electrical charges at regular intervals being measured in nanoseconds. (A nanosecond is a billionth of a second.) As Bolter remarks (p. 101) 'The clock has been at the center of Western technology since its invention in the Middle Ages. Computer technology too finds it indispensable, although it has changed the clock from a mechanical device to a wholly electronic one.'

Whenever we try to predict the future we are compelled to make our forecasts on the basis of what we believe to be the relevant aspects of current knowledge, although this means that we are largely guided by what has already happened. Consequently, when planning for tomorrow's world it is extremely difficult for us to cast off the dead hand of the past. It has been said that 'revolution is evolution in too much of a hurry', but the evolution of science and technology has become so rapid that it must itself be regarded as 'revolutionary'! The decision process must be continually speeded up if we are to innovate effectively and at the same time cope satisfactorily with the consequences of innovation.

As long ago as 1774, in a famous speech to the electors of Bristol, Edmund Burke maintained that their task was not to choose a mere spokesman for their own opinions but to elect a delegate endowed with the freedom to make his own decisions on the various issues to be debated in Parliament. In future, our rulers will also have to exhibit both foresight and some degree of specialized knowledge when deciding on the appointment of, and the duties to be assigned to, the professional systems-analysts who will be increasingly needed to assist government in ensuring that our economy functions successfully at home and at the same time is kept competitive in world-markets. Unlike pure scientists, who must have the necessary flair to decide which details have to be discarded and which retained in order to succeed in their investigations, the systems-analyst must be trained to examine *all* the aspects of a particular problem, including the likely effect on people, and must never be allowed to forget that 'systems' exist for people and not people for 'systems'. In the jargon associated with the use of modern computers, the term 'system' denotes the co-operative organization of the 'hardware' (i.e. the electronic and other material components of the computer), the 'software' (i.e. its programming) and the individuals needed to perform the tasks that must be undertaken to attain the desired results.

Frequently nowadays the use of highly sophisticated computers involves the 'selling' of time on them. In the Middle Ages this practice would have been severely frowned upon by the Church, for one of its main objections to the practice of usury was that it contravened natural law by 'selling time', and in its view time necessarily belongs to all creatures. According to the author of the *Tabula exemplorum*, written at the end of the thirteenth century (and according also to Duns Scotus, *c*.1260–1307, in his commentary on the *Sentences* of Peter Lombard), 'since usurers sell nothing but the hope of money, that is, time, they are selling the day and the night. But the day is the time of light and the night the time of rest; therefore, they are selling eternal light and rest.'[5]

Modern industrial society is dependent on time to a greater extent than any previous civilization, except possibly the Maya, but there is a peculiar ambivalence in this dependence. Whereas our knowledge of the past of both man and the universe is far greater than that possessed by our ancestors, our feeling of continuity with the past has tended to diminish owing to the rapid and continual changes that influence our lives. For many people today time has become so fragmented that only the present appears to be significant, the past being regarded as 'out-of-date' and therefore useless. Moreover, because the present differs so much from the past, it is becoming increasingly difficult to realize what the past was like. As Hans Meyerhoff has remarked, 'The past "is being ground to pieces" by the mill of inexorable, incomprehensible change.'[6]

Nevertheless, despite this drastic foreshortening of temporal perspective in our daily lives that makes the present seem all-important, the opposite influence prevails when, in trying to understand the nature of society and of the physical world, we believe that only by studying the past can we hope to understand the present. Consequently, nowadays the past is simultaneously devalued and enhanced in value.

This paradoxical situation is due to the dynamic nature of modern civilization. In the Middle Ages society was far more static and was essentially hierarchical in nature. As a result the causal or genetic attitude was far less important in medieval thought than it is in ours and the concept of evolution had little influence compared with the role of symbolism in the general world-view of the time. Moreover, even the concept of time itself was of less significance for historians then than it is for their successors today. We regard it as one of the first duties of the historian to date events precisely, the date being regarded by us not as an accidental property of an event but as an essential feature. This attitude, however, is comparatively modern. For St Augustine the date of an event was of far less importance than its theological significance. His tendency to see everything in a theological rather than in a historical perspective was a powerful influence in the Middle Ages, but during the Reformation, when papal tradition was under attack, historiography began to assume a new and strategic importance. Nevertheless, although the disputes of Protestants and Catholics stimulated much scholarly research on the past, for historians such as Bossuet the influence of divine providence was still the dominant factor, history being regarded as essentially a religious epic extending from the Creation to the Day of Judgement. Although in the sixteenth century a purely secular view of history had already been taken by Machiavelli and Guicciardini, it was not until the nineteenth century that the fundamental significance of historical perspective came to be generally recognized. This was several hundred years after the theory and practice of spatial perspective had been developed by painters and others. In each case a new way of looking at the world resulted. It was not surprising that history became a subject of major importance in its own right in the nineteenth century, for both the French revolution and the industrial revolution caused people to become far more conscious than previously of the reality and inevitability of change. They therefore felt the need to trace its progress.

As a result, the study of history was greatly encouraged. When he was Minister of Education in the French government in the 1830s Guizot arranged for the publication of a vast number of medieval chronicles at state expense. At the same time in Victorian Britain increased interest was shown in the past, particularly in the Middle Ages, whereas previously scholarly interest had been largely confined to Classical antiquity. In 1838 the Camden Society was formed for the recovery and examination of manuscripts. It was mainly in Germany, however, that the study of history made its greatest advances in the nineteenth century. British historians down to Lord Acton (1834–1902) and F. W. Maitland (1850–1906) repeatedly expressed their indebtedness to German scholarship, which was dominated by Leopold von Ranke (1795–1886) and Theodor Mommsen (1817–1903) and covered many fields including Classical antiquity and biblical studies. During the present century the scholarly study of history has been extended to all fields of knowledge, including the history of science, in which the United States has played a leading role.

One of the most striking manifestations this century of the greatly increased appreciation of the past and of our need to reconstruct it as far as possible from its surviving remains is the widespread interest in archaeology. It was not until the closing decades of the nineteenth century that the idea that excavations could be a useful non-literary means of adding to our knowledge of the past was introduced into Great Britain by General Pitt-Rivers. The introduction of carbon-dating and other sophisticated techniques this century· has been a powerful means of increasing the value of archaeology in this respect.

In many civilizations there has been an underlying analogy between the prevailing concepts of the nature of society and of the universe, and these analogies have often been associated with particular views of the nature and significance of time. For example, the Athenians of the sixth century BC regarded time as a judge. This was when the state was being founded on the concept of justice, and this concept was soon extended to explain the whole universe. Another example is provided by developments in the European Middle Ages and Renaissance when, following the invention of the mechanical clock, the idea of the mechanical simulation of the universe by clockwork suggested the reciprocal idea that the universe itself is a clocklike machine, an idea which came to the fore in the seventeenth century. The mechanistic analogy not only gave rise to the idea of the clocklike universe but also to a quasi-mechanical concept of human society that was most clearly described in the Introduction to Hobbes's *Leviathan*, of 1651, where the state is regarded as an artificial man and man himself is described mechanistically. Currently we have the historical analogy that originated in the eighteenth century and according to which both the universe and society are regarded as evolving in time.

Not only has the concept of change come to dominate our idea of human history, but in the last two centuries belief in the unchanging character of the physical universe has also been seriously undermined. Until the nineteenth century the concept of evolution made little impact on our way of thinking about the world. Astronomy, the oldest and most advanced science, did not indicate any evidence of trend in the universe. Although it had long been realized that time itself could be measured by the motion of the heavenly bodies and that the accuracy of man-made clocks could be controlled by reference to astronomical observations, the pattern of celestial motions, like that of a system of wheels, appeared to be the same whether read forwards or backwards, and the future was regarded as essentially a repetition of the past. Consequently, it was natural for people to lay primary emphasis on the cyclical aspects of time and the universe. When eventually they began to question the age-old belief that the overall state of the world remains more or less the same indefinitely, the concept of evolution was thought to characterize both living organisms and the physical world in general. As a result, the cyclical aspects of phenomena are now regarded as subordinate to long-term irreversibility.

It is a commonplace nowadays to regard everything as having a history and this applies even to our idea of time. The philosopher Immanuel Kant believed that the concept of time is a prior condition of our minds that affects our experience of the world, but this does not explain why different human societies have had different concepts of time and have assigned different degrees of significance to the temporal aspect of phenomena. It is now coming to be realized that, instead of being a prior condition, our concept of time should be regarded as a consequence of our experience of the world, the result of a long evolution. The human mind has the power, apparently not possessed by animals, to construct the idea of time from our awareness of certain features characterizing the data of our experience. Although Kant threw no light on the origin of this power, he realized that it was a peculiarity of the human mind. In recent years it has become clear that all our mental abilities are potential capacities which we can only realize in practice by learning how to use them. For, whereas animals inherit particular patterns of sensory awareness, known as 'releasers' because they automatically initiate certain types of action, humans have to learn to construct all their patterns of awareness from their own experience. Consequently, our ideas of space and time, which according to Kant function as if they were releasers, must instead be regarded as mental constructs that have to be learned.

The continuing evolution of our idea of time is revealed by the increasing importance of tense in the development of language. Greater knowledge of the universe has been accompanied by greater appreciation of the distinctions between past, present, and future as people have learned to transcend the limitations of 'the eternal present'. Although our awareness of time is based on psychological factors and on physiological processes below the level of consciousness, we have seen that it is also dependent on social and cultural influences. Because of these, there is a reciprocal relation between time and history. For, just as our idea of history is based on that of time, so time as we conceive it is a consequence of our history.

Appendix 1. Leap Years

According to Simon Newcomb, the tropical year at epoch AD 1900 is equivalent to 365.24219879 mean solar days, approximately.* Hence, to the nearest fifth decimal place, the fractional part of the number of days in the tropical year is 0.24220. This can be expanded as a simple continued fraction.†

$$0.24220 = \cfrac{1}{4 +} \ \cfrac{1}{7 +} \ \cfrac{1}{1 +} \ \cfrac{1}{3 +} \ \cdots\cdots ,$$

the first four convergents, i.e. successive approximations, being

$$\frac{1}{4}, \ \frac{7}{29}, \ \frac{8}{33}, \ \frac{31}{128},$$

respectively. The first convergent gives the Julian leap year rule, according to which every fourth year contains a leap day. The fourth convergent gives one fewer leap years in each period of 128 years, that is, 31 as against 32 Julian leap years, and would lead to an extremely accurate value for the average length of the calendar year, viz. 365.2421875 days, which is too short by about one second only. It is more convenient, however, to use the Gregorian calendar which gives 97 leap years in each period of 400 years, although it is less accurate, producing one too many leap years (776 instead of 775) in each period of 3,200 years. In fact, the Gregorian leap year rule gives the fractional part of the average number of days in the year as 0.2425 instead of 0.2422.

A somewhat more accurate approximation is given by the third

---

* The tropical year decreases by about 0.00006 days in 1000 years. When the Julian calendar was introduced (45 BC) it was approximately 365.24232 days.

† Simple continued fraction expansions tend to give much more accurate approximations than decimal expansions. The notation here used is more convenient than printing

$$\cfrac{1}{4 + \cfrac{1}{7 + \ldots}}$$

convergent above, viz. 8/33. It corresponds to the suggestion attributed to Omar Khayyam of eight leap years in each period of 33 years,* which yields the decimal approximation 0.24242 for the fractional part of the average number of days in the year. This rule would not be convenient to use, however, particularly because some leap years would occur in even-numbered years and some in odd-numbered ones.

If the Gregorian calendar were slightly modified, so that in addition to the present rules governing leap years all years divisible by 4,000 were taken to be ordinary years, there would be 969 (instead of 970) leap years in a period of 4,000 years, giving an average length of 365.24225 days for the calendar year. This is only about four seconds too long, corresponding to one day too many in about 20,000 years.

---

* The poet and mathematician Omar Khayyam was one of eight astronomers appointed, *c.* AD 1079, by the Sultan of Khorasan to reform the calendar.

Appendix 2. The Metonic Calendrical Cycle

Meton's cycle depended on the discovery that 235 synodic months or lunations (new moon to new moon) are very nearly equal to 19 tropical years (vernal equinox to vernal equinox). This can be easily checked, since the mean synodic month is about 29.5306 days and the tropical year, as we have seen in Appendix 1, is about 365.2422 days. The Metonic ratio can be attained by calculating the fifth convergent of the simple continued fraction for the decimal part of the number of months in the year. This gives the ratio as

$$12 + \cfrac{1}{2+} \; \cfrac{1}{1+} \; \cfrac{1}{2+} \; \cfrac{1}{1+} \; \cfrac{1}{1} = \frac{235}{19}$$

After 19 years the mean phases of the moon tend to recur on the same days of the month (with perhaps a shift of one day according to the number of leap years in the cycle) and within about two hours of their previous times. The months originally involved were 110 of 29 days and 125 of 30 days. The total number of days in the cycle was therefore 6,940, and consequently the average number of days in the year was a little in excess of 365.26. The particular cycle introduced by Meton began on the thirteenth day of the twelfth month of the calendar then used in Athens, which was 27 June 432 BC according to our reckoning. It appears that this day was chosen because Meton had determined astronomically that it was the summer solstice.

A more accurate version of Meton's cycle based on the assumption that the year is equal to 365.25 days was introduced about 330 BC by the astronomer Callippus, who found that Meton's 19-year cycle was slightly too long. He therefore combined four 19-year periods into one cycle of 76 years and dropped one day from the period, so that his cycle contained 27,759 days. Although it never came into general use, it became the standard for later astronomers and chronologists, for example Ptolemy. The number of days assigned to the year by Callippus became the basis of the Julian calendar.

Appendix 3. The Calculation of Easter

Unlike our civil calendar which is purely solar and the Islamic calendar which is purely lunar, the Christian ecclesiastical calendar depends on both the Sun and the Moon. Initially, the problem was complicated by differences between the various Christian Churches concerning the extent to which Jewish practices should be followed. Jewish law ordained that the Paschal Lamb must be slaughtered on the fourteenth day (beginning at nightfall) of Nisan, the first month of the ecclesiastical year, which began in the spring. According to the Gospels of Matthew, Mark, and Luke, since Christ was the true Paschal Lamb the Last Supper occurred on the day of the Jewish Paschal Feast, but according to John's Gospel that was the day of the crucifixion. A further complication was that the Jewish Feast could occur on any day of the week, whereas most Christians eventually wished the day of the Resurrection (two days after the crucifixion) to be on a Sunday. Only those in Asia Minor adhered to a definite date of the Jewish calendar, and as a result were called Quarto-decimans. This Paschal controversy first became a matter of general concern in the second century and led Polycarp, Bishop of Smyrna, to visit the Roman Pope Anicetus in the year 158. They agreed that each should adhere to his own practice. Forty years later a much more bitter controversy occurred between the Roman Pope Victor and Polycrates, Bishop of Ephesus, but eventually peace was restored by the Bishop of Lyons, Irenaeus.

Overshadowing these doctrinal differences, however, the determination of the relevant dates was complicated by the use of different methods of calculation, so that by the beginning of the fourth century important centres of Christianity such as Rome and Alexandria were celebrating Easter at very different times. At the request of the Emperor Constantine, the question was considered by the Council of Nicaea in the year 325. Unfortunately, the records we have of that Council are largely silent on this important issue, but later the same century Ambrose, Archbishop of Milan, in a letter that has survived, wrote that the Council had decreed that the western practice should prevail, so that Easter must be celebrated on the Sunday following the first full moon after the spring equinox. This Sunday was chosen so as to ensure that Easter never coincided with the Jewish Passover. The Quartodecimans refused to accept this decision and their practice continued in Asia Minor until the sixth century. The expression 'full moon' in connection with Easter means the ecclesiastical full moon, that is, the fourteenth day of the moon reckoned from the day of the first appearance of the moon after conjunction. The actual determination of this was referred to the astronomers of Alexandria, who alone were technically competent to deal with it.

According to Eusebius (*Church History*, vii. 32), Anatolius, Bishop of Laodicea, had already begun using the Metonic cycle for determining Easter *c*.277. This method was adopted in Alexandria, with the equinox taken to occur on 21 March, instead of 19 March as Anatolius had assumed. Eusebius mentions (*Church History*, vii. 20) that Bishop Dionysius of Alexandria had previously proposed a rule for Easter based on an 8-year cycle. This corresponds to the third convergent of the continued fraction for the number of months in the year, that is,

$$12 + \cfrac{1}{2 +} \ \cfrac{1}{1 +} \ \cfrac{1}{2}$$

from the formula in Appendix 2, which is equal to 99/8. It implies that there are approximately 99 lunations in 8 years. This is the octaeteris cycle referred to by Geminus (p. 45). Later, Victorius, Bishop of Aquitaine, introduced (*c*.457) a new cycle combining the Metonic cycle of 19 years with a solar cycle of 28 years (28 being the product of 7, the number of days in the week, and 4, the number of years in the leap year cycle) so as to produce a new cycle of 532 years for Easter. This came to be called the 'Dionysian period', because it was used by the Roman abbot Dionysius Exiguus in constructing the Easter tables that he calculated at the command of the emperor Justinian in the sixth century. Dionysius himself produced Easter tables only for the period 532 to 627, but later Isidore of Seville (*c*.560–636) continued them until 721. In the eighth century Bede completed this 532-year cycle by calculating tables down to the year 1063. The calculation of Easter was called the *computus*.

In the West regional differences in the dating of Easter ceased by the end of the eighth century, but by the thirteenth century the divergence of the spring equinox from 21 March began to be a cause for concern, since it then amounted to seven or eight days. This divergence was pointed out by, among others, Sacrobosco (John of Holywood, *fl.* 1230) in his *De anni ratione*, and by Roger Bacon (*c*.1219–92) in his *De reformatione calendaris*, transmitted to the Pope. Nevertheless, it was not until 1474 that Pope Sixtus IV invited the leading astronomer of the day,

Regiomontanus, to Rome for the reconstruction of the calendar. His premature death delayed further action, and so it was not until 1582 that the more accurate Gregorian calendar replaced the Julian calendar.

The Julian calendar was based on the inaccurate assumption that the tropical year is exactly 365.25 days. The other inaccurate assumption that affected the determination of Easter and the Church calendar was that, according to the Metonic cycle, 235 lunations are exactly equal to 19 Julian years. By 1582 the error in the lunar cycle from this cause amounted to about four days so that the fourteenth day of the Church calendar moon was the eighteenth day of the actual mean moon. A method of calculation was suggested by Aloisius Lilius which involved abandoning the Metonic cycle and replacing the Golden Number by the Epact. The term 'Golden Number' was coined to indicate the place which any year occupies in the Metonic cycle, that is, the age of the moon on a given date, because the Greeks are said to have inscribed these numbers in gold on public pillars. For years AD the rule for obtaining this Number is to add one to the number of the particular year concerned, for example 1582, and find the remainder on dividing by 19, with the additional proviso that when this remainder is zero the Golden Number is taken to be 19. Because the Golden Numbers were only adapted to the Julian calendar, Lilius in his proposed reform of that calendar used Epacts instead, an 'Epact' being the whole number denoting the lunar phase, that is the age of the calendrical moon, on a definite date, for example, 1 January. Following this method, the Papal astronomer Christopher Clavius computed new tables for the determination of Easter according to the Gregorian calendar.

Nowadays it is not necessary to appeal to the Clavius tables in order to determine the date of Easter, because in 1800 an elegant mathematical formula for this purpose was devised by the great German mathematician Carl Friedrich Gauss (1777–1855). Previously, a set of mathematical rules of the same general character was devised by Thomas Harriot (1560–1621) but was never published. (Like much of Harriot's scientific work, it has only come to light in recent years.) Gauss's rule, when applied to any year in the present century written as $1900 + N$, can be stated as follows:

(1) calculate the remainders a, b, and c, when N is divided by 19, 4, and 7, respectively;

(2) calculate the remainder d, when $19a + 24$ is divided by 30;

(3) calculate the remainder e, when $2b + 4(c - 1) + 6d$ is divided by 7;

(4) if the sum $d + e$ does not exceed 9, Easter occurs on the day in March given by $d + e + 22$, but if $d + e$ exceeds 9, Easter occurs on the day in April given by $d + e - 9$.

For example, for the year 1988, we have $\dot{N} = 88$, $a = 12$, $b = 0$, $c = 4$, $d = 12$, $e = 0$, and hence Easter falls on 3 April.

Unfortunately, Gauss's neat solution fails to give correct results for some years after 4200, and so in 1817 the problem was further investigated by the French astronomer Jean-Baptiste Delambre (1749–1822). Sixty years later a thorough re-examination of the problem was made by Samuel Butcher, Bishop of Meath.[1] In 1876 a New York correspondent had sent, without proof, to the weekly scientific periodical *Nature* a rule for the determination of Easter which, unlike Gauss's rule, is subject to no exceptions.[2] Butcher showed that this rule followed from Delambre's analytical solution. For any given year n the rule is as follows:

Divide	By	Quotient	Remainder
n	19		a
n	100	b	c
b	4	d	e
$b + 8$	25	f	
$b - f + 1$	3	g	
$19a + b - d - g + 15$	30		h
c	4	i	k
$32 + 2e + 2i - h - k$	7		l
$a + 11h + 22l$	451	m	
$h + l - 7m + 114$	31	p	q

The number of the month in which Easter falls is given by p and the day of the month by $q + 1$. For example, for the year 1988 this calculation yields $p = 4$, $q = 2$, and hence we conclude that Easter Day that year is 3 April. The earliest day on which Easter can occur is 22 March and the latest 25 April. Uspensky and Heaslet provide an elementary mathematical discussion of calendrical problems, including the calculation of Easter.[3]

References

Preface

1   G. J. Whitrow, *The Natural Philosophy of Time* (London and Edinburgh; Nelson, 1961; Oxford: Clarendon Press, 1961; 2nd edn., 1980).
2   S. Toulmin and J. Goodfield, *The Discovery of Time* (London: Hutchinson, 1965; Harmondsworth: Penguin Books, 1967).
3   R. Wendorff, *Zeit und Kultur: Geschichte des Zeitbewusstseins in Europa* (Wiesbaden: Westdeutscher Verlag, 1980).
4   H. Trivers, *The Rhythm of Being: A Study of Temporality* (New York: Philosophical Library, 1985), part III: 'Time and History'.

Chapter 1   Awareness of Time

1   R. Wallis, *Le Temps, quatrieme dimension de l'esprit* (Paris: Flammarion, 1966), 51 ff.
2   J. Piaget, *The Child's Conception of Time*, trans. A. J. Pomerans (London: Routledge & Kegan Paul, 1969).
3   A. E. Wessmann and B. S. Gorman, *The Personal Experience of Time* (New York: Plenum Press, 1977), 8.
4   E. Michaud, *Essai sur l'organisation de la connaissance entre 10 et 14 ans* (Paris: Vrin, 1949).
5   P. M. Bell, 'Sense of time', *New Scientist* (15 May 1975), 406.
6   C. Ralling, 'A vanishing race', *Listener* (16 July 1959), 87.
7   W. Koehler, *The Mentality of Apes*, trans. from 2nd rev. edn. Ella Winter (Harmondsworth: Penguin Books, 1957), 234.
8   S. Walker, *Animal Thought* (London: Routledge & Kegan Paul, 1983), 190.
9   B. L. Whorf, *Language, Thought and Reality*, ed. J. B. Carroll (Cambridge, Mass.: MIT Press, 1956), 57–64.
10   Ibid. 58.
11   S. C. McCluskey, 'The astronomy of the Hopi Indians', *Journal for the History of Astronomy*, 8 (1977), 174–95.
12   E. E. Evans-Pritchard, *Witchcraft, Oracles and Magic among the Azande* (Oxford: Clarendon Press, 1937), 347.
13   E. E. Evans-Pritchard, *The Nuer: A Description of the Modes of Livelihood and Political Institutions of a Nilotic People* (Oxford: Clarendon Press, 1940), 103.
14   Ibid. 105.
15   Ibid. 108.

Chapter 2 Describing Time

1 E. H. Lenneberg, *Biological Foundations of Language* (New York: Wiley, 1968), 106.

2 C. M. Bowra, 'Some aspects of speech', in *In General and Particular* (London: Weidenfeld & Nicolson, 1966), 14.

3 R. E. Passingham, 'Broca's area and the origin of human vocal skill', *Phil. Trans. Roy. Soc. (London)*, B292 (1981), 167–75.

4 G. Steiner, *After Babel: Aspects of Language and Translation* (Oxford University Press, 1975), 157.

5 S. Fleischman, *The Future in Thought and Language* (Cambridge University Press, 1982), 50.

6 Steiner, op. cit. (above, n. 4), 139.

7 Whitrow, *Natural Philosophy of Time* (2nd edn.; preface, n. 1), 174 ff.

8 M. P. Nilsson, *Primitive Time-reckoning* (Lund: C. W. K. Gleerup, 1920), 9–10.

Chapter 3 Time at the Dawn of History

1 P. Radin, *Primitive Man as Philosopher*, 2nd edn. (New York: Dover, 1957).

2 Ibid. 244.

3 A. Marshack, 'Some implications of the Palaeolithic symbolic evidence for the origins of language', *Current Anthropology*, 17 (1976), 274.

4 R. S. Solecki, 'Shanidar IV, a Neanderthal flower burial in northern Iraq', *Science*, 190 (1975), 880.

5 D. C. Heggie, *Megalithic Science: Ancient Mathematics and Astronomy in North-west Europe* (London: Thames & Hudson, 1981).

6 S. G. F. Brandon, *Time and Mankind* London: Hutchinson, 1951), 33.

7 H. Frankfort *et al.*, *Before Philosophy* (Harmondsworth: Penguin Books, 1949), 35.

8 O. Neugebauer, *The Exact Sciences in Antiquity* (Providence, RI: Brown University Press, 1957), 81.

9 H. E. Winlock, 'The origin of the ancient Egyptian calendar', *Proc. Amer. Phil. Soc.*, 83 (1940), 447.

10 J. H. Breasted, 'The beginnings of time-measurement and the origins of our calendar', in *Time and its Mysteries*, Series I (New York University Press, 1936), 80.

11 T. G. H. James, *An Introduction to Ancient Egypt* (London: British Museum Publications, 1979), 125.

12 Neugebauer, loc. cit. (above, n. 8).

13 S. N. Kramer, *The Sumerians* (Chicago and London: University of Chicago Press, 1963), 328.

14 J. G. Gunnell, *Political Philosophy and Time* (Middleton, Conn.: Wesleyan University Press, 1968), 40.

15   E. Voegelin, *The Ecumenic Age* (vol. 4 of *Order and History*) (Baton Rouge: Louisiana State University Press, 1980), 84.

16   G. Contenau, *Everyday Life in Babylon and Assyria*, trans. K. R. and A. R. Maxwell-Hyslop (London: Edward Arnold, 1954), 213.

17   N. K. Sanders, *The Epic of Gilgamesh* (Harmondsworth: Penguin Books, 1960), 104.

18   D. Pingree, 'Astrology', in P. P. Wiener (ed.), *Dictionary of the History of Ideas* (New York: Scribner, 1973), i. 118.

19   A. Sachs, 'Babylonian horoscopes', *Journal of Cuneiform Studies*, 6 (1952), 49.

20   Seneca, *Nat. Quaest.* III. 29. 1 (London: Heinemann, 1971), 286.

21   O. Neugebauer, 'The history of ancient astronomy: problems and methods', *Publications of the Astronomical Society of the Pacific*, 58 (1946), no. 340, 33.

22   O. Neugebauer, *A History of Ancient Mathematical Astronomy* (Berlin: Springer Verlag, 1975), i. 4.

23   Ibid. ii. 593.

24   R. C. Zaehner, *Dawn and Twilight of Zoroastrianism* (London: Weidenfeld & Nicolson, 1961), 55.

25   R. C. Zaehner, *Zurvan: A Zoroastrian Dilemma* (Oxford: Clarendon Press, 1955), 410.

26   S. G. F. Brandon, *Creation Legends of the Ancient Near East* (London: Hodder & Stoughton, 1963), 206.

27   W. Hartner, 'The Young-Avestan and Babylonian calendars and the antecedents of precession', *Journal for the History of Astronomy*, 10 (1979), 1–22.

28   S. H. Taqizadah, *Old Iranian Calendars* (London: Royal Asiatic Society, 1938).

29   E. Yarshater, 'Time-reckoning', in *Cambridge History of Iran* (Cambridge University Press, 1982), ii. 790.

Chapter 4   *Time in Classical Antiquity*

1   Gunnell, op. cit. (ch. 3, n. 14), 15.

2   F. M. Cornford, *From Religion to Philosophy* (London: Edward Arnold, 1912), 181.

3   W. K. C. Guthrie, 'The religion and mythology of the Greeks', in *The Cambridge Ancient History*, rev. edn. (Cambridge University Press, 1961), ii, ch. 40, 39–40.

4   H. Lloyd Jones, *The Justice of Zeus* (Berkeley: University of California Press, 1971), 5–6 and 166–7 n. 23.

5   W. Jaeger, *The Theology of the Early Greek Philosophers*, trans. E. S. Robinson (London: Oxford University Press, 1967), 35.

6   Whitrow, *Natural Philosophy of Time* (ch. 2, n. 7), 190–200.

7   Nemesius, Bishop of Emesa, in E. Bevan, *Later Greek Religion* (London: Dent, 1927), 30–1.

8   L. Edelstein, *The Idea of Progress in Late Antiquity* (Baltimore: Johns Hopkins University Press, 1967), xxi.

9   R. Drews, *The Greek Accounts of Eastern History* (Cambridge, Mass.: Harvard University Press, 1973), 35–6.

10 M. I. Finley, 'Thucydides the moralist', in *Aspects of Antiquity* (Harmondsworth: Penguin Books, 1977), 53.

11 A. Momigliano, 'The place of Herodotus in the history of historiography', in *Studies in Historiography* (London: Weidenfeld & Nicolson, 1966), 130.

12 J. de Romilly, *Time in Greek Tragedy* (Ithaca: Cornell University Press, 1968), 5–6.

13 E. R. Dodds, 'Progress in classical antiquity', in P. P. Wiener (ed.), *Dictionary of the History of Ideas* (New York: Scribner, 1973), iii. 633.

14 A. Momigliano, 'Time in ancient historiography', in *History and Theory*, 1966, Suppl. 6 ('History and the concept of time'), 10.

15 W. K. C. Guthrie, *In the Beginning: Some Greek Views on the Origin of Life and the Early State of Man* (London: Methuen, 1957), 65.

16 Momigliano, 'Time in ancient historiography' (above, n. 14), 13.

17 P. Duhem, *Le Système du monde* (Paris: Hermann, 1954), ii (new edn.), 299.

18 Alexander of Aphrodisias, *On Destiny: Addressed to the Emperors*, trans. A. Fitzgerald (London: Scholaris Press, 1931), 25.

19 Kramer, op. cit. (ch. 3, n. 13), 262.

20 W. K. C. Guthrie, *A History of Greek Philosophy* (Cambridge University Press, 1969), iii. 82.

21 Guthrie, *In the Beginning* (above, n. 15), 79.

22 Guthrie, *A History of Greek Philosophy* (above, n. 20), iii. 292.

23 J. V. Noble and D. J. de Solla Price, 'The water-clock in the Tower of Winds', *Amer. J. Archaeol.*, 72 (1968), 345–55.

24 T. C. Vriezen, *The Religion of Ancient Israel*, trans. H. Hoskins (London: Lutterworth Press, 1969), 243.

25 O. Cullmann, *Christ and Time*, trans. F. V. Filson (London: SCM Press, 1951), 51.

26 Gunnell, op. cit. (ch. 3, n. 14), 75.

27 G. W. Trompf, *The Idea of Historical Recurrence in Western Thought* (Berkeley: University of California Press, 1979), 134.

28 W. O. E. Oesterley, *The Evolution of the Messianic Idea* (London: Isaac Pitman & Sons, 1908), 206.

29 Gunnell, op. cit. (ch. 3, n. 14), 63–4.

30 H. Frankfort, *Kingship and the Gods: A Study of Near Eastern Religion and the Integration of Society and Nature* (Chicago: University of Chicago Press, 1978; Phoenix edn.), 343–4.

31 E. Voegelin, *Israel and Revelation* (vol. 1 of *Order and History*) (Baton Rouge: Louisiana State University Press, 1956).

32 G. Van Seters, *In Search of History: Historiography in the Ancient World and*

the Origins of Biblical History (New Haven and London: Yale University Press, 1983), 241.

33   H. Webster, *Rest Days: A Study in Early Law and Morality* (New York: Macmillan, 1916), 252.

34   Ibid. 254.

35   Vriezen, op. cit. (above, n. 24), 234.

36   M. Testuz, *Les Idées religieuses du Livre des Jubilées* (Geneva: Droz; Paris: Minard, 1960), 136.

37   L. Casson, *Travel in the Ancient World* (London: Allen & Unwin, 1974), 155.

38   R. Syme, *The Roman Revolution* (London: Oxford University Press, 1960), 315–16.

39   J. T. Shotwell, *The History of History* (New York: Columbia University Press, 1939), 301.

40   E. R. Curtius, *European Literature and the Latin Middle Ages*, trans. W. R. Trask (New York: Pantheon Books, 1953), 252.

41   Lucretius, *The Nature of the Universe*, trans. R. E. Latham (Harmondsworth: Penguin Books, 1951), 40–1.

42   P. Brown, *The World of Late Antiquity: From Marcus Aurelius to Muhammad* (London: Thames & Hudson, 1971), 62.

43   Ibid.

44   H.-C. Puech, 'Gnosis and time', in *Man and Time: Papers from the Eranos Yearbooks* (London: Routledge & Kegan Paul, 1958), 61.

45   F. Cumont, *The Mysteries of Mithra*, trans. T. J. McCormack (Chicago: Open Court, 1903), 1.

46   Ibid. 39.

47   M. J. Vermaseren, 'A magical time god', in J. R. Hinnells (ed.), *Mithraic Studies: Proceedings of the First International Congress of Mithraic Studies, 1971* (Manchester University Press, 1975), 451.

48   E. A. Wallis Budge, *Osiris and the Egyptian Resurrection* (London: Philip Lee Warner, 1911), i. 60.

49   Vermaseren, op. cit. (above, n. 47), 456.

50   S. Sambursky and S. Pines, *The Concept of Time in Late Neoplatonism* (Jerusalem: Israel Academy of Sciences and Humanities, 1971), 11.

51   J. F. Callahan, *Four Views of Time in Ancient Philosophy* (Cambridge, Mass.: Harvard University Press, 1948), 124.

52   C. N. Cochrane, *Christianity and Classical Culture: A Study of Thought and Action from Augustus to Augustine* (London: Oxford University Press, 1974), 186.

53   E. Frank, *Philosophical Understanding and Religious Truth* (New York: Oxford University Press, 1945), 68.

54   J. Baillie, *The Belief in Progress* (Cambridge University Press, 1951), 76.

55   O. Pedersen, 'The ecclesiastical calendar and the life of the Church', in G. V. Coyne, M. A. Hoskin, and O. Pedersen (eds.), *Gregorian Reform of*

the Calendar (Vatican City: Pontifica Academica Scientiarum, 1983), 22.

56 Frank, op. cit. (above, n. 53), 70.

57 R. L. Poole, 'The beginning of the year in the middle ages', in *Studies in Chronology and History* (Oxford: Clarendon Press, 1934), 1–27.

58 E. J. Bickerman, *Chronology of the Ancient World* (London: Thames & Hudson, 1968), 77.

59 F. K. Ginzel, *Handbuch der Chronologie*, vol. iii (Leipzig: Hinrichs, 1914), 115.

60 F. H. Colson, *The Week: An Essay on the Origin and Development of the Seven-day Cycle* (Cambridge University Press, 1926).

61 Bickerman, op. cit. (above, n. 58), 61.

62 H. I. Marrou, *A History of Education in Antiquity*, trans. G. Lamb (London: Sheed & Ward, 1956), 148.

63 Cochrane, op. cit. (above, n. 52), 330–1.

64 G. Teres, 'Time computations and Dionysius Exiguus', *Journal for the History of Astronomy*, 15 (1984), 177–88.

Chapter 5 Time in the Middle Ages

1 R. W. Southern, *Medieval Humanism and Other Studies* (Oxford: Blackwell, 1970), 3.

2 M. L. W. Laistner, 'The library of the Venerable Bede', in A. Hamilton Thompson (ed.), *Bede, His Life, Times, and Writings: Essays in Commemoration of the Twelfth Centenary of his Death* (Oxford: Clarendon Press, 1935), 238.

3 A. Bryant, *A History of Britain and the British People*, vol. 1: *Set in a Silver Sea* (London: Collins, 1984), 29.

4 Bede, *The Ecclesiastical History of the English Nation* (London: Dent, 1935; Everyman edn.), 152.

5 R. L. Poole, 'Imperial influences on the forms of Papal documents', in *Studies in Chronology and History* (Oxford: Clarendon Press, 1934), 178.

6 J. A. Burrow, *The Ages of Man: A Study in Medieval Writing and Thought* (Oxford: Clarendon Press, 1986), 29–30.

7 Southern, op. cit. (above, n. 1), 158.

8 Ibid. 162.

9 C. H. Haskins, *Studies in the History of Medieval Science*, 2nd edn. (Cambridge, Mass.: Harvard University Press, 1927), 117; see also Southern, op. cit. (above, n. 1), 166–7; and Bodleian MS. Auct.F.1.9, fo. 90.

10 W. Hartner, 'The principle and use of the astrolabe', in *Oriens–Occidens* (Hildesheim: Georg Olms, 1968), 287–318; J. D. North, 'The astrolabe', *Scientific American*, 230 (Jan. 1974), 96–106.

11 D. J. de Solla Price, 'Mechanical water clocks of the 14th century in Fez, Morocco', in *Proceedings of the Tenth International Congress of the History of Science (Ithaca, 1962)* (Paris: Hermann, 1964), i. 599–602.

12   D. R. Hill (ed. and trans.), *On the Construction of Water-clocks* (London: Turner & Devereux, 1976), 9.

13   D. R. Hill (ed. and trans.), *The Book of Ingenious Devices* (Dordrecht: Reidel, 1974), 271 ff.

14   D. B. MacDonald, 'Continuous re-creation and atomic time in Muslim scholastic theology', *Isis*, 9 (1927), 326–7.

15   M. Maimonides, *The Guide for the Perplexed*, trans. A. Friedlander (London: Routledge, 1904), 121.

16   MacDonald, op. cit. (above, n. 14), 341.

17   al-Biruni, *The Chronology of Ancient Nations*, trans. and ed. E. C. Sachau (London: W. H. Allen, 1879), 34–6.

18   L. Massignon, 'Time in Islamic thought', in *Man and Time: Papers from the Eranos Yearbooks* (London: Routledge & Kegan Paul, 1958), 109.

19   B. Smalley, *Historians of the Middle Ages* (London: Thames & Hudson, 1974), 30.

20   A. J. Gurevich, *Categories of Medieval Culture*, trans. G. L. Campbell (London: Routledge & Kegan Paul, 1985), 122.

21   N. Cohn, *The Pursuit of the Millenium* (London: Secker & Warburg, 1957), 102.

22   M. Reeves, *Joachim of Fiore and the Prophetic Future* (London: SPCK, 1976), 3.

23   M. Reeves, *The Influence of Prophecy in the Later Middle Ages* (Oxford: Clarendon Press, 1969), 296.

24   R. Garaudy, 'Faith and revolution', *Ecumenical Review*, 25 (1973), 66–7.

25   R. S. Westfall, *Never at Rest: A Biography of Isaac Newton* (Cambridge University Press, 1980), 319 ff.

26   M. Bloch, *Feudal Society*, trans. L. A. Manyon (London: Routledge & Kegan Paul, 1961), 73.

27   Ibid. 74.

28   J. U. Nef, *Cultural Foundations of Industrial Civilizations* (Cambridge University Press, 1958), 17.

29   R. Glasser, *Time in French Life and Thought*, trans. C. G. Pearson (Manchester University Press, 1972), 17.

30   Ibid. 56.

31   R. Pernoud, *Joan of Arc*, trans. E. Hyams (Harmondsworth: Penguin Books, 1969), 31.

32   A. Murray, *Reason and Society in the Middle Ages* (Oxford: Clarendon Press, 1985), 107.

33   R. L. Poole, *Medieval Reckonings of Time* (London: SPCK, 1918), 46–7.

34   J. Gairdner, *The Paston Letters 1422–1509 AD. Introduction and Supplement* (Westminster: Archibald Constable, 1901), p. ccclxvi.

35   R. J. Quinones, *The Renaissance Discovery of Time* (Cambridge, Mass.: Harvard University Press, 1973), 110.

36   Ibid. 113.

37   L. White, *Medieval Technology and Social Change* (Oxford: Clarendon Press, 1962), 61.

Chapter 6   Time in the Far East and Mesoamerica

1   H. Jacobi, 'Atomic theory (Indian)' in *Dictionary of Religion and Ethics* (Edinburgh: Clark, 1909, ii. 202.

2   A. N. Balslev, *A Study of Time in Indian Philosophy* (Wiesbaden: Otto Harrassowitz, 1983), 39 ff.

3   M. Eliade, *Images and Symbols: Studies in Religious Symbolism*, trans. P. Mairet (London: Harvill Press, 1961), 65.

4   J. Needham and Wang Ling, *Science and Civilisation in China* (Cambridge University Press, 1959), iii. 315.

5   J. Needham, Wang Ling, and D. J. de Solla Price, *Heavenly Clockwork: The Great Astronomical Clocks of Medieval China* (Cambridge University Press, 1960).

6   F. A. B. Ward, 'How timekeeping became accurate', *Chartered Mechanical Engineer*, 8 (1961), 604.

7   S. A. Bedini, 'The scent of time: a study of the use of fire and incense for time measurement in oriental countries', *Trans. Amer. Phil. Soc.*, 53 (1963), Part 5, 6.

8   J. H. Plumb, *The Death of the Past* (London: Macmillan, 1969), 111.

9   J. Needham, 'Time and knowledge in China and the West', in J. T. Fraser (ed.), *The Voices of Time* (New York: Braziller, 1966), 96.

10   J. Needham, *Time and Eastern Man* (Henry Myers Lecture) (London: Royal Anthropological Institute, 1965), Occasional Paper no. 21, 8–9.

11   V. H. Malmstrom, 'Origin of the Mesoamerican 260-day calendar', *Science*, 181 (1973), 939–41.

12   R. J. Wenke, *Patterns in Prehistory* (New York: Oxford University Press, 1984), 383.

13   N. Hammond, *Ancient Maya Civilization* (Cambridge University Press, 1982), 199 ff.

14   J. E. S. Thompson, *A Commentary on the Dresden Codex: A Maya Hieroglyphic Book* (Philadelphia: American Philosophical Society, 1972), 62–70.

15   M. Leon-Portilla, *Time and Reality in the Thought of the Maya*, trans. C. L. Boiles and F. Horcasitas (Boston: Beacon Press, 1973), 91–2.

16   J. E. S. Thompson, *The Rise and Fall of Maya Civilization* (London: Gollancz, 1956), 145.

17   S. G. Morley, *The Ancient Maya* (Stanford, Calif.: Stanford University Press, 1947), 449.

18   D. S. Landes, *Revolution in Time* (Cambridge, Mass.: Harvard University Press, 1983), 24.

Chapter 7 The Advent of the Mechanical Clock

1 D. J. de Solla Price, 'Gears from the Greeks: the Antikythera mechanism—a calendar computer from ca. 80 BC', *Trans. Amer. Phil. Soc.*, 64 (1974), Part 7, 1–70.

2 J. V. Field and M. T. Wright, 'Gears from the Byzantines: a portable sundial with calendrical gearing', *Annals of Science*, 42 (1985), 87.

3 L. White, *Medieval Technology and Social Change* (Oxford: Clarendon Press, 1962), 120.

4 E. Panofsky, *Studies in Iconology* (Oxford: Clarendon Press, 1939), 80.

5 L. Thorndike, 'Invention of the mechanical .clock about 1271 AD', *Speculum*, 16 (1941), 242–3.

6 C. F. C. Beeson, *English Church Clocks 1280–1850* (London and Chichester: Phillimore (Antiquarian Horological Society), 1971), 13.

7 J. D. North, 'Monasticism and the first mechanical clocks', in J. T. Fraser and N. Lawrence (eds.), *The Study of Time*, ii (Berlin: Springer Verlag, 1975), 385.

8 J. D. North, *Richard of Wallingford* (Oxford: Clarendon Press, 1976), i. 441–526.

9 A. J. Dudeley, *The Mechanical Clock of Salisbury Cathedral* (Salisbury: Friends of Salisbury Cathedral Publishing, 1973).

10 White, op. cit. (above, n. 3), 124–5.

11 S. A. Bedini and F. R. Maddison, 'Mechanical universe: the Astrarium of Giovanni de' Dondi', *Trans. Amer. Phil. Soc.*, 56 (1966), Part 5, 60.

12 J. Le Goff, *Time, Work and Culture in the Middle Ages*, trans. A. Goldhammer (Chicago: University of Chicago Press, 1980), 46.

13 J. Harthan, *Books of Hours and Their Owners* (London: Thames & Hudson, 1977), 39.

14 F. Hattinger, *The Duc de Berry's Book of Hours* (Berne: Hallwag, 1970).

15 J. Huizinga, *The Waning of the Middle Ages*, trans. F. Hopman (Harmondsworth: Penguin Books, 1972), 149–50.

16 K. Thomas, *Religion and the Decline of Magic* (London: Weidenfeld & Nicolson, 1971), 621.

17 L. Mumford, *Technics and Civilization* (London: Routledge & Kegan Paul, 1934), 14.

18 I. Origo, *The Merchant of Prato* (London: Jonathan Cape, 1957), 177.

19 H. Tait, *Clocks and Watches* (London: British Museum Publications, 1983), 43.

20 Landes, op. cit. (ch. 6, n. 18), 89.

21 J. Aubrey, *Brief Lives and Other Selected Writings*, ed. A. Powell (London: Cresset Press, 1949), 133.

22 F. M. Powicke and A. B. Emden, *The Universities of Europe in the Middle Ages* (Oxford University Press, 1936), iii. 401.

23 A. Palmer, *Movable Feasts: Changes in English Eating-habits* (Oxford University Press, 1984).
24 F. Rabelais, *Gargantua* (1535), i. 23.

Chapter 8 Time and History in the Renaissance and the Scientific Revolution

1 L. Pastor, *History of the Popes*, ed. R. F. Kerr, Vol. 19 (London: Kegan Paul, 1930), 293.
2 H. M. Nobis, 'The reaction of astronomers to the Gregorian calendar', in G. V. Coyne, M. A. Hoskin, and O. Pedersen (eds.), *Gregorian Reform of the Calendar* (Vatican City: Pontifica Academia Scientiarum, 1983), 250.
3 R. M. Dawkins, *The Monks of Athos* (London: Allen & Unwin, 1936), 198.
4 J. M. Thompson, *Leaders of the French Revolution* (Oxford: Blackwell, 1948), 159.
5 H. Webster, *Rest Days* (New York: Macmillan, 1916), 283.
6 C. Cipolla, *Clocks and Culture: 1300–1700* (London: Collins, 1967), 42.
7 J. Drummond Robertson, *The Evolution of Clockwork* (London: Cassell, 1931), 54–61.
8 E. Grant, *Nicole Oresme and the Kinematics of Circular Motion* (Madison: University of Wisconsin Press, 1971), 295.
9 R. Boyle, *The Works of the Honourable Robert Boyle*, ed. T. Birch (London: 1772), v. 163.
10 A. R. Hall, 'Horology and criticism: Robert Hooke', *Studia Copernicana*, XVI, Ossolineum, 1978, 261–81.
11 Mumford, op. cit. (ch. 7, n. 17), 15.
12 I. Barrow, *Lectiones Geometricae*, trans. E. Stone (London: 1735), Lecture 1, 35.
13 G. W. Leibniz, *Philosophical Writings*, trans. M. M. (London: Dent, 1934), 200.
14 R. Boyle, *The Excellence of Theology Compared with Natural Philosophy*, 1665 (London: 1772), 11.
15 E. Breisach, *Historiography: Ancient, Medieval and Modern* (Chicago: University of Chicago Press, 1983), 177.
16 F. Manuel, *Isaac Newton Historian* (Cambridge University Press, 1963), 274.
17 C. Morris, *The Tudors* (Glasgow: Fontana-Collins, 1966), 12.
18 G. J. Whitrow, *What is Time?* (London: Thames & Hudson, 1972), 19–20.
19 A. Kent Hieatt, *Short Time's Endless Monument: The Symbolism of the Numbers in Edmund Spenser's 'Epithalamion'* (Port Washington, NY, and London: Kennikat Press, 1972), 81.

20   R. W. Hepburn, 'Cosmic fall', in P. P. Wiener (ed.), *Dictionary of the History of Ideas* (New York: Scribner, 1968), i. 505–6.

21   D. Seward, *The First Bourbon: Henry IV of France and Navarre* (London: Constable, 1971), 133.

22   M. Tiles, 'Mathesis and the masculine birth of time', *International Studies in the Philosophy of Science*, 1 (1986), 16–35.

23   F. Saxl, 'Veritas filia temporis', in R. Klibansky and H. J. Paton (eds.), *Philosophy and History: The Ernst Cassirer Festschrift* (Oxford: Clarendon Press, 1936). Reprinted as Harper Torchbook (Harper & Row, 1963), 197–222.

24   R. V. Sampson, *Progress in the Age of Reason: The Seventeenth Century to the Present Day* (London: Heinemann, 1956), 99.

25   C. Becker, *The Heavenly City of the Eighteenth Century Philosophers* (New Haven: Yale University Press, 1968; 1st edn. 1932), 130.

26   F. Smith Fussner, *The Historical Revolution: English Historical Writing and Thought 1580–1640* (London: Routledge & Kegan Paul, 1962), 166.

27   E. L. Eisenstein, 'Clio and Chronos' in *History and Theory*, 1966, Suppl. 6 ('History and the concept of time'), 47.

Chapter 9   *Time and History in the Eighteenth Century*

1   R. W. Symonds, *Thomas Tompion: His Life and Work* (London: Batsford, 1951), 10.

2   *Journals of the House of Commons*, 11 June 1714, 677.

3   J. Swift, *Gulliver's Travels* (London: Dent, 1940; Everyman edn.), 224.

4   H. Quill, *John Harrison: The Man Who Found Longitude* (London: John Baker, 1966), 59.

5   R. T. Gould, *The Marine Chronometer: Its History and Development* (London: Potter, 1923), 50 ff.

6   Quill, op. cit. (above, n. 4), 317.

7   R. T. Gould, *John Harrison and his Timekeepers* (London: National Maritime Museum, 1958), 12.

8   Gould, *Marine Chronometer* (above, n. 5), 86.

9   G. W. Leibniz, *The Monadology and other Philosophical Writings*, trans. R. Latta (London: Oxford University Press, 1925), 350–1.

10   A. O. Lovejoy, *The Great Chain of Being* (Cambridge, Mass.: Harvard University Press, 1948), 246.

11   R. Nisbet, *History of the Idea of Progress* (London: Heinemann, 1980), 180.

12   Sampson, op. cit. (ch. 8, n. 24), 240.

13   E. Cassirer, *Rousseau, Kant, Goethe* (Princeton University Press, 1945), 56.

14   M. J. Temmer, *Time in Rousseau and Kant* (Geneva: Droz and Paris: Minard, 1958), 31.

15   R. Haynes, *Philosopher King: The Humanist Pope Benedict XIV* (Weidenfeld & Nicolson, 1970), 178.

16   I. Berlin, *Vico and Herder* (London: Hogarth Press, 1976), 142 n.

17 Ibid. 38.
18 R. G. Collingwood, *The Idea of History* (Oxford: Clarendon Press, 1948), 68.
19 Berlin, op. cit. (above, n. 16), 143 ff.
20 G. J. Whitrow, *Kant's Cosmogony*, trans. W. Hastie (New York and London: Johnson Reprint Corp., 1970), xi–xl.
21 S. Toulmin and J. Goodfield, *The Discovery of Time* (Harmondsworth: Penguin Books, 1967), 167.
22 Berlin, op. cit. (above, n. 16), 150–1.

Chapter 10   Evolution and the Industrial Revolution

1   W. Herschel, *Phil. Trans. Roy. Soc.* (1814), 284.
2   Lovejoy, op. cit. (ch. 9, n. 10), 243.
3   N. Hampson, *The Enlightenment* (Harmondsworth: Penguin Books, 1968), 220.
4   R. Taton (ed.), *The Beginning of Modern Science*, trans. A. J. Pomerans (London: Thames & Hudson, 1964), 572–3.
5   A. Geikie, *The Founders of Geology* (London: Macmillan, 1897), 283.
6   J. D. Burchfield, *Lord Kelvin and the Age of the Earth* (London: Macmillan, 1975), 136–40.
7   J. Perry, 'On the age of the earth', *Nature*, 51 (3 Jan. 1895), 224–7. See also his letter on the same topic (18 Apr. in the same vol.), 582–5.
8   G. H. Darwin, *The Tides* (London: Murray, 1898), 257.
9   P. Burke, *The Renaissance Sense of the Past* (London: Edward Arnold, 1969), 141.
10 L. Wright, *Clockwork Man* (London: Elek, 1968), 128.
11 F. Klemm, *A History of Western Technology*, trans. D. W. Singer (Cambridge, Mass.; MIT Press, 1964), 196.
12 Wright, op. cit. (above, n. 10), 128.
13 J. Simmons, *The Railway in England and Wales 1830–1914*, vol. 1 (Leicester University Press, 1978), 23.
14 Wright, op. cit. (above, n. 10), 143.
15 J. A. Bennett, 'George Biddell Airy and horology', *Annals of Science*, 37 (1980), 268–85.
16 Wright, op. cit. (above, n. 10), 147.
17 Mumford, op. cit. (ch. 7, n. 17), 14.
18 Ibid. 17.
19 E. Gellner, *Times Literary Supplement* (23 Dec. 1983), 1, 438.
20 D. Howse, *Greenwich Time and the Discovery of Longitude* (Oxford University Press, 1980), 113–14.
21 S. Kern, *The Culture of Time and Space: 1880–1918* (London: Weidenfeld & Nicolson, 1983), 12.
22 L. Essen, *The Measurement of Frequency and Time Interval* (London: HMSO, 1973).

Chapter 11 Rival Concepts of Time

1 G. Poulet, *Studies in Human Time*, trans. E. Coleman (New York: Harper, 1959), 200.
2 Whitrow, *Natural Philosophy of Time* (Preface n. 1), 103 ff.
3 G. J. P. Scrope, *The Geology and Extinct Volcanoes of Central France* (London: John Murray, 1858), 208.

Chapter 12 Civilization as Progress?

1 A. de·Toqueville, *Democracy in America*, trans. H. Reeve (London: Oxford University Press, 1946), 311.
2 G. Stent, *Paradoxes of Progress* (San Francisco: Freeman, 1978), 27.
3 W. Thomson, Lord Kelvin, 'Nineteenth-century clouds over the theory of heat and light', in *Baltimore Lectures on Molecular Dynamics and the Wave Theory of Light* (Cambridge University Press, 1904), Appendix B, 486–527.
4 J. David Bolter, *Turing's Man: Western Culture in the Computer Age* (London: Duckworth, 1984).
5 Le Goff, op. cit. (ch. 7, n. 12), 290.
6 H. Meyerhoff, *Time in Literature* (Berkeley and Los Angeles: University of California Press, 1955), 109.

Appendices

1 S. Butcher, *The Ecclesiastical Calendar: Its Theory and Construction* (Dublin: Hodges, Foster & Figgis: London: Macmillan, 1877).
2 Anon., *Nature*, 13 (1876), 487.
3 J. V. Uspensky and M. A. Heaslet, *Elementary Number Theory* (New York and London: McGraw-Hill, 1939), 206–21.

Index